Semi-analytic Function, Conjugate Analytic Function and Their Tremendous Influences

Wang Jianding

中国水利水电出版社
China Water & Power Press
· 北京 ·

ABSTRACT

This book has comprehensively generalized analytical functions, proposed a new concept of semi-analytical functions, conjugate analytical functions and corresponding theories, and explained the application of this theory in electromagnetic fields, fluid mechanics, elastic mechanics, geometric transformation and other fields. The theory is relatively systematic and the content is novel. Because it provides a method for how to propose new concepts and how to apply new concepts, it is an important reference book for scientists engaged in theoretical research and applied development research. This book can also be used as a learning material for complex variable functions in science and engineering colleges.

图书在版编目（CIP）数据

半解析函数、共轭解析函数及其重大影响 = Semi-analytic Function, Conjugate Analytic Function and Their Tremendous Influences : 英文 / 王见定著. -- 北京 : 中国水利水电出版社, 2018.12
ISBN 978-7-5170-7288-1

Ⅰ. ①半… Ⅱ. ①王… Ⅲ. ①解析函数－研究－英文 Ⅳ. ①O174.55

中国版本图书馆CIP数据核字(2018)第295830号

书　　名	**Semi-analytic Function, Conjugate Analytic Function and Their Tremendous Influences**
中文书名拼音	BANJIEXI HANSHU, GONG'EJIEXI HANSHU JI QI ZHONGDA YINGXIANG
作　　者	Wang Jianding（王见定）著
出版发行	中国水利水电出版社 （北京市海淀区玉渊潭南路1号D座　100038） 网址：www.waterpub.com.cn E-mail：sales@waterpub.com.cn 电话：(010) 68367658（营销中心）
经　　售	北京科水图书销售中心（零售） 电话：(010) 88383994、63202643、68545874 全国各地新华书店和相关出版物销售网点
排　　版	中国水利水电出版社微机排版中心
印　　刷	天津嘉恒印务有限公司
规　　格	170mm×230mm　16开本　5印张　106千字
版　　次	2018年12月第1版　2018年12月第1次印刷
印　　数	0001—1000册
定　　价	**68.00**元

PREFACE

In 1983, Prof. Wang Jianding was the first person in the world to make advance the semi-analytic function and its application in mechanics. In 1988, he systematically established the conjugate analytic function theory and made these two theories successfully to have their applications in the areas of electric, magnetic, fluid mechanics, elasticity mechanics. These two theories have been quoted and developed worldwide by lots of specialists and scholars, then many new branches of math were founded and rapidly developed, such as bianalytic function, complex harmonious function, k analytic function, semi-bianalytic function, semi-conjugate analytic function and corresponding the boundary value problem. The theory of Prof. Wang Jianding's semi-analytic function and conjugate analytic function is the extension and development of the theory of analytic function which was created by the world great masters of math: Cauchy, Riemann, Weierstrass, Guasss and Euler.

CONTENTS

PART Ⅲ Their Tremendous Influences

PART Ⅰ Semi-analytic Function

Chapter Ⅰ Definitions and Existence Theorems

1.1 Definitions

Let $f(z)=u(x,y)+iv(x,y)$ be continuous in domain D. It is assumed that following $f(z)$ is continuous.

Definition 1.1 A $f(z)$ is called semi-analytic of the first kind at point (x, y), if u_x and v_y are continuous in a neighborhood of (x, y), and $u_x=v_y$ at the point (x, y).

Definition 1.2 A $f(z)$ is called semi-analytic of the first kind in domain D, if $f(z)$ is semi-analytic of the first kind for every point (x, y) in D.

Definition 1.3 A $f(z)$ is called semi-analytic of the second kind at point (x, y), if u_y and v_x are continuous in a neighborhood of (x, y), and $u_y=-v_x$ at the point (x, y).

Definition 1.4 A $f(z)$ is called semi-analytic of the second kind in domain D. if $f(z)$ is semi-analytic of the second kind for every point (x, y) in D.

Definition 1.5 Let $f(z)$ be semi-analytic in D, we consider a greater domain G (than D) which contains D. If $F(z)$ is semi-analytic in G and $F(z)=f(z)$ in D, then $F(z)$ is called a semi-analytic extension of $f(z)$ in G.

Definition 1.6 $\{D, f(z)\}$ is called a semi-analytic element, where D is a domain and $f(z)$ is semi-analytic in D. $\{D_1, F_1(z)\}=\{D_2, f_2(z)\}$ if and only if $D_1=D_2$, $f_1(z)=f_2(z)$.

1.2 Existence Theorems

Theorem 1.1 If $f(z)=u(x,y)+iv(x,y)$ is analytic in D, then $f^*(z)=[u(x,y)+\varphi(y)]+i[v(x,y)+\psi(x)]$ is semi-analytic of the first kind in D, where $\varphi(y)$ and $\psi(x)$ are arbitrary continuous functions in D.

Theorem 1.2 If $f(z)=u(x,y)+iv(x,y)$ is analytic in D, then $f^*(z)=[u(x,y)+\psi(x)]+i[v(x,y)+\varphi(y)]$ is semi-analytic of the second kind in D, where $\varphi(y)$ and $\psi(x)$ are arbitrary continuous functions in D.

Chapter Ⅱ Principal Properties

2.1 Algebraic and Analytic Properties

Theorem 2.1 If $f_1(z)$ and $f_2(z)$ are semi-analytic of the first (second) kind, then $a_1 f_1(z)+a_2 f_2(z)$ is semi-analytic of the first (second) kind, where a_1 and a_2 are real numbers.

Theorem 2.2 If $f(z)$ is semi-analytic of the first kind in D, u_y and v_x are continuous and $\frac{f(z)}{z-z_0}$ is semi-analytic of the first kind in D (except for z_0), then $f(z)$ is analytic in D.

Theorem 2.3 If $f(z)$ is semi-analytic of the second kind, u_x and v_y are contiunous in D and $\frac{f(z)}{z-z_0}$ is semi-analytic of the second kind in D (except for z_0), then $f(z)$ is analytic in D.

It is assumed that following contours are all rectifiable Jordan-curves.

Theorem 2.4 Let $f_n(z)=u_n+iv_n$ $(n=1, 2, \cdots)$ be semi-analytic of the first kind in D.

If

$$\begin{aligned} &(1)\ \lim_{n\to\infty} f_n(z)=f(z) \\ &(2)\ \left\{\frac{\partial u_n}{\partial x}\right\} \text{ is convergent uniformly} \end{aligned} \tag{1.2.1}$$

then $f(z)$ is semi-analytic of the first kind in D.

Theorem 2.5 Let $f_n(z)=u_n+iv_n$ $(n=1, 2, \cdots)$ be semi-analytic of the second kind in D.

If

$$\begin{aligned} &(1)\ \lim_{n\to\infty} f_n(z)=f(z) \\ &(2)\ \left\{\frac{\partial u_n}{\partial y}\right\} \text{ is convergent uniformly} \end{aligned} \tag{1.2.2}$$

then $f(z)$ is semi-analytic of the second kind in D.

Theorem 2. 6 If $f(z)$ is semi-analytic of the first kind in simply connected domain D and C is an arbitrary contour in D, then $\int_C f(z)dz$ is a real number.

The extension of theorem 2. 6 is as follows:

Theorem 2. 7 Let C be a contour, D be interior of C and $G=C+D$. If $f(z)$ is semi-amalytic of the first kind in D and continuous on G, then $\int_C f(z)dz$ is a real number.

Theorem 2. 8 If $f(z)$ is semi-analytic of the second kind in simly connected domain D, C is an arbitrary contour in D, then $\int_C f(z)dz$ is a imaginary number.

The extension of theorem 2. 8 is as follows:

Theorem 2. 9 Let C be a contour, D be interior of C and $G=C+D$. If $f(z)$ is semi-analytic of the second kind in D and continuous on G, then $\int_C f(z)dz$ is an imaginary number.

Theorem 2. 10 Let D be a complex connected domain and $C=C_0+C_1+\cdots+C_n$ be bound of D (C_i is contour, $i=0, 1, \cdots, n$). If $f(z)$ is semi-analytic of the first kind in D, then $\int_C f(z)dz$ is a real number.

Theorem 2. 11 Let D be a simply connected domain and let u_x and v_y be continuous in D. If $\int_C f(z)dz$ is a real number for arbitrary contour $C\subset D$, then $f(z)$ is semi-analytic of the first kind in D.

Theorem 2. 12 Let D be a simply connected domain and let u_y and v_x be continuous in D. If $\int_C f(z)dz$ is an imaginary number for arbitrary contour $C\subset D$, then $f(z)$ is semi-analytic of the second kind in D.

It is obvious that theorem 2. 11 and theorem 2. 12 also hold for complex connected domain.

Theorem 2. 13 If $f(z)$ is semi-analytic of the first (second) kind in

D, then $if(z)$ is semi-analytic of the second (first) kind in D.

Theorem 2. 13 expresses $90°$-rotation of plane vecter in geometry.

Theorem 2. 14 If $f(z)$ is semi-analytic of the second kind, then exists surely $\varphi(x, y)$ such that $\overline{f(z)} = \nabla\varphi(x, y)$, and such $\varphi(x, y)$ is innumerable, but the difference between each other is a constant. On the contrary, if $\overline{f(z)} = \nabla\varphi(x, y)$, then $f(z)$ is surely semi-analytic of the second kind.

Theorem 2. 15 If $f(z)$ is semi-analytic of the first kind, then exists surely $\varphi(x, y)$ such that $\overline{if(z)} = \nabla\varphi(x, y)$, and such $\varphi(x, y)$ is innumerable, but the difference between each other is a constant. On the contrary, if $\overline{if(z)} = \nabla\varphi(x, y)$, then $f(z)$ is surely semi-analytic of the first kind, where $\nabla\varphi(x,y) \equiv \left(\frac{\partial\varphi}{\partial x}, \frac{\partial\varphi}{\partial y}\right)$, $\varphi(x, y)$ and its first, second-order partial derivatives are continuous.

Theorem 2. 16 If $f(z) = \big(u(x,y), v(x,y)\big)$ is semi-analytic of the first kind, then exists $\mathbf{b} = (b_x, b_y, b_z)$ such that

$$\mathbf{a} = \big(u(x,y), -v(x,y), w(x,y)\big) = \nabla \times \mathbf{b}$$

where $w(x, y)$ and its first-order, second-order partial derivatives are continuous. On the contrary, if $\mathbf{a} = \big(u(x,y), -v(x,y), w(x,y)\big) = \nabla \times \mathbf{b}$, then $f(z) = \big(u(x,y), v(x,y)\big)$ is surely semi-analytic of the first kind. $\mathbf{b}$ satisfy $\mathbf{a} = \nabla \times \mathbf{b}$ is innumerable, but the difference between each other is $\nabla\varphi$.

Theorem 2. 17 If $f(z) = \big(u(x,y), v(x,y)\big)$ is semi-analytic of the second kind, then exists $\mathbf{b} = (b_x, b_y, b_z)$ such that

$$\mathbf{a} = \big(-v(x,y), -u(x,y), w(x,y)\big) = \nabla \times \mathbf{b}$$

On the contrary , if $\mathbf{a} = \big(-v(x,y), -u(x,y), w(x,y)\big) = \nabla \times \mathbf{b}$, then $f(z) = \big(u(x,y), v(x,y)\big)$ is surely semi-analytic of the second kind. $\mathbf{b}$ satisfy $\mathbf{a} = \nabla \times \mathbf{b}$ is innumerable, but the difference between each other is $\nabla\varphi$, where φ is an arbitrary scalar-valued function, $w(x, y)$ and its first, second-order partial derivatives are continuous.

In the following, we consider only semi-analytic function of the first kind.

Theorem 2.18 Let $\{D_1, f_1(z)\}$ and $\{D_2, f_2(z)\}$ be two semi-analytic elements, if

$$\begin{aligned}&(1)\ D_1 \cap D_2 = d_{12}, D_1 \not\supset D_2 \text{ and } D_2 \not\supset D_1\\&(2)\ f_1(z) = f_2(z), z \in d_{12}\end{aligned} \tag{1.2.3}$$

then $\{D_1 + D_2, F(z)\}$ is also a semi-analytic element, where

$$F(z) = \begin{cases} f_1(z), & z \in D_1 - d_{12} \\ f_1(z) = f_2(z), & z \in d_{12} \\ f_2(z), & z \in D_2 - d_{12} \end{cases} \tag{1.2.4}$$

Theorem 2.19 Let $\{D_1, f_1(z)\}$ and $\{D_2, f_2(z)\}$ be two semi-analytic elements, if

(1) $D_1 \cap D_2 = \Phi$, and open arc Γ is common bound of D_1 and D_2.

$$(2)\ \begin{matrix} f_1(z), u_x^{(1)} \text{ and } v_y^{(1)} \text{ are continuous in } D_1 + \Gamma \\ f_2(z), u_x^{(2)} \text{ and } v_y^{(2)} \text{ are continuous in } D_2 + \Gamma \end{matrix}$$

$$(3)\ f_1(z) = f_2(z) \quad z \in \Gamma$$

then $\{D_1 + \Gamma + D_2, F(z)\}$ is also a semi-analytic element, where

$$F(z) = \begin{cases} f_1(z), & z \in D_1 \\ f_1(z) = f_2(z), & z \in \Gamma \\ f_2(z), & z \in D_2 \end{cases} \tag{1.2.5}$$

For semi-analytic function of the second kind we also have similar conclusions.

Theorem 2.20 Let D and D^* be two domains in z-plane; they locate respectively in upper semi-plane and below semi-plane and are symmetric with regard to X-axis; segment $S \subset X$-axis is a part of their bound. Let $\{D+S, f(z)\}$ be a semi-analytic element of the first kind (second kind), $f(z)$ be continuous in $D+S$, and $f(z)$ be real on S, then

$$F(z) = \begin{cases} f(z), & z \in D + S \\ \overline{f(\bar{z})}, & z \in D^* \end{cases} \tag{1.2.6}$$

is a semi-analytic of the first kind (second kind) in $D+S+D^*$.

Remark: $F(z)$ may be multiform in extended domain.

2.2 Decomposition Theorems of Complex Functions

Theorem 2.21 Let partial derivetives of first order of $f(z)$ be Hölder continuous in domain D, then $f(z)$ can be decomposed into a sum of two semi-analytic functions. That is,

$$f(z)=f_1(z)+f_2(z) \tag{1.2.7}$$

where $f_1(z)$ and $f_2(z)$ are respectively semi-analytic of the first kind and the second kind in D.

Theorem 2.22 Let first and second order partial derivatives of $f(z)$ be continuous on closed domain D, then $f(z)$ can be decomposed into a sum of two semi-analytic functions. That is,

$$f(z)=f_1(z)+f_2(z) \tag{1.2.8}$$

where $f_1(z)$ and $f_2(z)$ are respectively semi-analytic of the first and second kind in D.

Chapter Ⅲ Physical Background and Applications

3.1 Physical Background

In the following, we will point out the physical background of semi-analytic function with the example of hydrodynamics.

Let the flow be steady in domain D of z-plane, $f(z)=u(x,y)+iv(x,y)$ be velocity of fluid at $z\in D$, $u(x,y)$, $v(x,y)$ and their first-order partial derivatives be continuous in D.

$\overline{f(z)}$ is called complex velocity.

Theorem 3.1 Steady flow is indivergent in $D \Leftarrow\Rightarrow \overline{f(z)}$ is semi-analytic of the first kind.

Theorem 3.2 Steady flow is irrotational in $D \Leftarrow\Rightarrow \overline{f(z)}$ is semi-analytic of the second kind.

It follows that all properties of indivergent (or irrotational) field can be transplanted into semi-analytic functions, and all properties of semi-analytic function can also be transplanted into indivegent (or irrotational) fields.

3.2 Applications of Semi-analytic Functions

Theorem 3.3 Indivergent field and irrotational field can mutually be translated by means of $90°$-rotation.

Theorem 3.4 If a rotated-$90°$ indivergent (irrotational) field is still indivergent (irrotational), then the field must be harmonic. This is a simple and convenient judgement method of harmonic field.

Theorem 3.5 Any plane-field, first order partial derivatives of which are Hölder continuous in finite domain D, (which is continuously partially differentiable for 2 times on closed domain) can be decomposed into an in-divergent plane-field and an irrotational plane-field.

3.3 Basic Function of Complex Function

Semi-analytic function is basic function of complerc function.

Theorem 3.6 Suppose G is a bunded planar domain, function

$$f(z)\in C'(G)\cap C(\overline{G}),\ \text{then}$$
$$f(z)=f_1(z)+f_2(z) \tag{1.3.1}$$

here $f_1(z)$, $f_2(z)$ are first class semi-analytic function and second class semi-analytic function respectly. and equation (1.3.1) is unique if we lost sight of the addition of arbitrary function $\Phi(z)$.

The conclusion expresses that a differentiable planar field can be regar-ded as a sum of a unsourece planar field and a uncurl planar field.

3.4 Integrul Format of Semi-analytic Function

Theorem 3.7 The function $w(z)=\iint\limits_D \frac{\overline{g(z_0)}}{\overline{z_0 - z}} \mathrm{d}\sigma_{z_0}$ is a second class semi-analytic function in D, here $g(z_0)$ is an arbitrary continuous real value function in D.

Theorem 3.8 The funtion $w(z) = i\iint\limits_D \frac{\overline{g(z_0)}}{\overline{z_0 - z}} \mathrm{d}\sigma_{z_0}$ is a first class semi-analytic function in D, here $g(z_0)$ is an arbitrary continuous real value function.

Above theorems have given complex integrul formats of uncurl planar field and unsource planar field, and a complex analytic tool have been pro-vided by them.

Chapter Ⅳ Boundary Value Problem of Semi-analytic Function

We consider nonhomogeneous C-R equation:

$$\begin{cases}\dfrac{\partial u}{\partial x}-\dfrac{\partial v}{\partial y}=p(x,y)\\[2mm]\dfrac{\partial u}{\partial y}+\dfrac{\partial v}{\partial x}=0\end{cases}$$

here $p(x, y)$ is an arbitrary hamonic function. Its complex form is

$$\frac{\partial w}{\partial \bar{z}}=p(x,y),\quad w=u+iv$$

whose general solution will be

$$\begin{aligned}w(z)&=\frac{-1}{\pi}\iint_G\frac{p(x,y)}{\zeta-z}\mathrm{d}\xi\mathrm{d}\eta+\Phi(z)\\&=\frac{-1}{\pi}\iint_G\frac{\varphi(\zeta)}{\zeta-z}\mathrm{d}\xi\mathrm{d}\eta+\frac{-1}{\pi}\iint_G\frac{\overline{\varphi(\zeta)}}{\zeta-z}\mathrm{d}\xi\mathrm{d}\eta+\Phi(z)\end{aligned}\tag{1.4.1}$$

where $\varphi(z)$ is an arbitrary analytic function,

$$w_1(z)=\frac{-1}{\pi}\iint_G\frac{\varphi(\zeta)}{\zeta-z}\mathrm{d}\xi\mathrm{d}\eta\tag{1.4.2}$$

is a bianalytic function and

$$w_2(z)=\frac{-1}{\pi}\iint_G\frac{\overline{\varphi(\zeta)}}{\zeta-z}\mathrm{d}\xi\mathrm{d}\eta\tag{1.4.3}$$

is a complex harmonic function.

Because of the uniqueness of bianalytic function and complex harmonic function, corresponding boundary value problem of semi-analytic function is unique.

The conclusion shows that we can alrea dy consider the phsical fields with sources or curls.

PART Ⅱ Conjugate Analytic Function

Chapter Ⅰ Conjugate Analytic Function

1.1 Definitions

Let function $w=f(z)$ be definite in domain D, $z\in D$, and $z+\Delta z\in D(\Delta z=\Delta x+i\Delta y)$. Then we compute the increatment of $f(z)$ at the z

$$\Delta w=f(z+\Delta z)-f(z)$$

If $\lim\limits_{\Delta z\to 0}\frac{\Delta w}{\overline{\Delta z}}$ is existent as $\Delta z\to 0$ in any manner, then $\lim\limits_{\Delta z\to 0}\frac{\Delta w}{\overline{\Delta z}}$ is called a conjugate derivative of $f(z)$ at z, and denoted by $f^{\circ}(z)$, that is,

$$f^{\circ}(z)=\lim_{\Delta z\to 0}\frac{\Delta w}{\overline{\Delta z}} \tag{2.1.1}$$

Right now we call that the function $f(z)$ is conjugate derived or conjugate differentiable at z.

Definition 1.1 If $w=f(z)$ is conjugate derived at every point in D, then $w=f(z)$ is called a conjugate analytic function in D or conjugate analytic in D.

The following conclusions are obvious.

$$(f_1(z)+f_2(z))^{\circ}=f_1^{\circ}(z)+f_2^{\circ}(z) \tag{2.1.2}$$

$$(f_1(z)f_2(z))^{\circ}=f_1^{\circ}(z)f_2(z)+f_1(z)f_2^{\circ}(z) \tag{2.1.3}$$

$$\left(\frac{f_1(z)}{f_2(z)}\right)^{\circ}=\frac{f_1^{\circ}(z)f_2(z)-f_1(z)f_2^{\circ}(z)}{[f_2(z)]^2}\quad [f_2(z)\neq 0] \tag{2.1.4}$$

Let $\xi=f(z)$ be conjugate analytic in D and $w=g(\overline{\xi})$ be conjugate analytic in G. If $\overline{\xi}=\overline{f(z)}$ belongs to G for every point in D, then $w=g(\overline{f(z)})$ is conjugate analytic in D, and

$$(g(\overline{f(z)}))^{\circ}=g^{\circ}(\overline{\xi})\cdot f^{\circ}(z) \tag{2.1.5}$$

Let $w=f(z)$ be conjugate analytic in D, $f^{\circ}(z)\neq 0$, $z\in D$, and P be a set $w=f(z)$. If $z=\psi(w)$ is a inverse function of $f(z)$, then $\psi(w)$ is conjugate analytic in P, and

$$\psi^{\circ}(w)=\frac{1}{\overline{f^{\circ}(\psi(w))}} \tag{2.1.6}$$

The equation systems

$$\begin{aligned} u_x &= -v_y \\ u_y &= v_x \end{aligned} \tag{2.1.7}$$

is called a condition of conjugate analytsis, where u_x, u_y, v_x and v_y are all continuous in D.

Theorem 1.1 Let $f(z)=u(x,y)+iv(x,y)$ be definite in D and be conjugate defferentiable at point $z\in D$. Then $u(x, y)$ and $v(x, y)$ have partial derivatives u_x, u_y, v_x and v_y at z, and they satisfy the condition of conjugate analysis.

Theorem 1.2 Let $f(z)=u(x,y)+iv(x,y)$ be definite in D. $f(z)$ is conjugate analytic at $z\in D$ if and only if $u(x, y)$ and $v(x, y)$ are differentiable at z and $u(x, y)$ and $v(x, y)$ satisfy the condition of conjugate analysis at z.

When the above conditions can be satisfied, $f^{\circ}(z)$ may be expressed by one of the following forms

$$f^{\circ}(z)=u_x+iv_x=-v_y+iv_x=u_x+iu_y=-v_y+iu_y \tag{2.1.8}$$

Theorem 1.3 $f(z)=u(x,y)+iv(x,y)$ is conjugate analytic in D if and only if u_x, u_y, v_x and v_y are continuous in D and $f(z)$ satisfies condition of conjugate analysis for every point in D.

Corollary 1.1 If $f(z)$ is conjugate analytic in D, and $f^{\circ}(z)=0(z\in D)$, then $f(z)=$constant in D.

Corollary 1.2 If $f(z)=u(x,y)+iv(x,y)$ is conjugate analytic in D, and $f^{\circ}(z)\neq 0$, then

$$u(x,y)=C_1 \text{ and } v(x,y)=C_2 \quad (C_1 \text{ and } C_2 \text{ are real constants}) \tag{2.1.9}$$

are two familis of curves which are orthogonal.

1.2 Conjugate Analytic Function and Harmonic Function

In this section we will discuss the relation between conjugate analytic function and harmonic function.

Definition 1.2 Let $f(z)=u(x,y)+iv(x,y)$ satisfy condition of conjugate analysis, and $u(x, y)$ and $v(x, y)$ be harmonic in D. Then $v(x, y)$ is called a conjugate harmonic function of $u(x, y)$.

Theorem 1.4 If $f(z)=u(x,y)+iv(x,y)$ is conjugate analytic in D, then $v(x, y)$ is surely a conjugate harmonic function of $u(x, y)$.

Theorem 1.5 Let $u(x, y)$ be a harmonic function in D, which be a simply connected domain. Then there is a $v(x, y)$ such that $f(z)=u(x,y)+iv(x,y)$ is conjugate analytic in D.

1.3 Conjugate Analytic Function and Analytic Function

In this section we will discuss relation between conjugate analytic and analytic function.

Theorem 1.6 If $f(z)$ is conjugate analytic in D, then $\overline{f(z)}$ is analytic in D.

Theorem 1.7 If $f(z)$ is analytic in D, then $\overline{f(z)}$ is conjugate analytic in D.

Theorem 1.8 A $f(z)$ is conjugate analytic in D if and only if $\overline{f(z)}$ is analytic in D.

Theorem 1.9 Let $f(z)$ be conjugate derived at point z. Then

$$f^{\circ}(z)=\overline{([\overline{f(z)}]')} \quad \text{at } z \tag{2.1.10}$$

Theorem 1.10 If $f(z)$ is conjugate derived at z, then the $f(z)$ has conjugate derivatives of arbitrary order at z.

$$f^{[n]}(z)=\overline{[\overline{f(z)}]^{(n)}} \quad (n=1,2,\cdots) \tag{2.1.11}$$

Note: $f^{[n]}(z)$ expresses the n-th order conjugate derivative of $f(z)$ at z.

1.4 Elementary Conjugate Analytic Functions

In this section some elementary conjugate analytic functions will be definited.

1.4.1 Exponential function

$$\overline{e^z}=e^x(\cos y-i\sin y) \text{ for arbitrary } z=x+iy. \tag{2.1.12}$$

Exponential function $\overline{e^z}$ has properties as follows.

$$|\overline{e^z}|=e^x>0 \quad \arg\overline{e^z}=-y \tag{2.1.13}$$

$$\overline{e^{z_1+z_2}}=\overline{e^{z_1}}\cdot\overline{e^{z_2}} \quad \text{for arbitrary } z_1 \text{ and } z_2. \tag{2.1.14}$$

$$\overline{e^{z+2\pi i}}=\overline{e^z} \tag{2.1.15}$$

$$(\overline{e^z})^\circ=\overline{e^z} \tag{2.1.16}$$

1.4.2 Logarithmic function

$$\begin{aligned} w=\overline{\mathrm{Ln}z}&=\ln|z|-i\mathrm{Arg}z\\ &=\ln|z|-i\arg z-2\pi ki \end{aligned} \tag{2.1.17}$$

(k is an arbitrary integer)

where $\qquad 0\leqslant\arg z<2\pi$

It follows that $w=\overline{\mathrm{Ln}z}$ has infinite values.

$$\overline{(\mathrm{Ln}z)_k}=\ln|z|-i\arg z-2\pi ki \tag{2.1.18}$$

It is called the k-th branch of $\overline{\mathrm{Ln}z}$ and is conjugate differentiable, and

$$[\overline{(\mathrm{Ln}z)_k}]^\circ=(\ln|z|-i\arg z-2\pi ki)^\circ=\frac{1}{\bar{z}} \tag{2.1.19}$$

The $w=\overline{\mathrm{Ln}z}$ has properties as follows:

$$\overline{\mathrm{Ln}(z_1\cdot z_2)}=\overline{\mathrm{Ln}z_1}+\overline{\mathrm{Ln}z_2} \tag{2.1.20}$$

$$\overline{\mathrm{Ln}\left(\frac{z_1}{z_2}\right)}=\overline{\mathrm{Ln}z_1}-\overline{\mathrm{Ln}z_2} \tag{2.1.21}$$

1.4.3 Power function

$$\overline{z^\alpha}=\overline{e^{\alpha\mathrm{Ln}z}} \tag{2.1.22}$$

where α is a constant and $z \neq 0$.

In general $\overline{z^\alpha}$ is multiform because $\mathrm{Ln}z$ is multiform.

Power function $\overline{z^\alpha}$ has properties as follows:

$$\overline{z^\alpha} = \overline{z^n}, \text{ when } \alpha = n \tag{2.1.23}$$

(n is a positive integer)

$$\overline{z^\alpha} = \frac{1}{\overline{z^n}}, \text{ when } \alpha = -n \tag{2.1.24}$$

$$\overline{z^\alpha} = \overline{z^{\frac{1}{n}}} = \frac{1}{\sqrt[n]{z}}, \text{ when } \alpha = \frac{1}{n} \tag{2.1.25}$$

$$(\overline{z^\alpha})^\circ = \overline{\alpha z^{\alpha - 1}} \tag{2.1.26}$$

Trigonometric functions

$$\overline{\sin z} \triangleq \overline{\left(\frac{e^{iz} - e^{-iz}}{2i}\right)} \tag{2.1.27}$$

$$\overline{\cos z} \triangleq \overline{\left(\frac{e^{iz} + e^{-iz}}{2}\right)} \tag{2.1.28}$$

$$\overline{\tan z} \triangleq \frac{\overline{\sin z}}{\overline{\cos z}} \tag{2.1.29}$$

They have the following important characteristics:

$$\overline{\cos z} - i\,\overline{\sin z} = \overline{e^{iz}} \tag{2.1.30}$$

$$\overline{\cos(-z)} = \overline{\cos z},\ \overline{\sin(-z)} = -\overline{\sin z} \tag{2.1.31}$$

$$(\overline{\sin z})^2 + (\overline{\cos z})^2 = 1 \tag{2.1.32}$$

$$\overline{\sin(z + 2\pi)} = \overline{\sin z},\ \overline{\cos(z + 2\pi)} = \overline{\cos z},\ \overline{\tan(z + \pi)} = \overline{\tan z} \tag{2.1.33}$$

$$\begin{aligned} \overline{\cos(z_1 + z_2)} &= \overline{\cos z_1} \cdot \overline{\cos z_2} - \overline{\sin z_1} \cdot \overline{\sin z_2} \\ \overline{\sin(z_1 + z_2)} &= \overline{\sin z_1} \cdot \overline{\cos z_2} + \overline{\cos z_1} \cdot \overline{\sin z_2} \end{aligned} \tag{2.1.34}$$

$$\begin{aligned} &(\overline{\sin z})^\circ = \overline{\cos z},\ (\overline{\cos z})^\circ = -\overline{\sin z} \\ &(\overline{\tan z})^\circ = -\frac{1}{(\overline{\cos z})^2} \end{aligned} \tag{2.1.35}$$

Likewise, we may definite

$$\overline{\mathrm{ctg} z} \triangleq \frac{\overline{\cos z}}{\overline{\sin z}},\ \overline{\sec z} \triangleq \frac{1}{\overline{\cos z}},\ \overline{\csc z} \triangleq \frac{1}{\overline{\sin z}} \tag{2.1.36}$$

Thus,

$$(\overline{\text{ctg}z})^{\circ}=-(\overline{\csc})^2,\ (\overline{\sec z})^{\circ}=\overline{\sec z}\cdot\overline{\tan z},\ (\overline{\csc z})^{\circ}=-\overline{\csc z}\cdot\overline{\tan z} \tag{2.1.37}$$

1.4.4 Inverse trigometric functions

$$\text{Arc}\ \overline{\sin z}=\frac{1}{i}\text{Ln}(i\,\bar{z}+\sqrt{1-(\bar{z})^2}) \tag{2.1.38}$$

$$\text{Arc}\ \overline{\cos z}=\frac{1}{i}\text{Ln}(\bar{z}+i\sqrt{1-(\bar{z})^2}) \tag{2.1.39}$$

$$\text{Arc}\ \overline{\tan z}=\frac{1}{2i}\text{Ln}\,\frac{1+i\,\bar{z}}{1-i\,\bar{z}} \tag{2.1.40}$$

1.4.5 Hyperbolic functions

$$\overline{\text{sh}z}\overset{\Delta}{=}\overline{\left(\frac{e^{z}-e^{-z}}{2}\right)},\ \overline{\text{ch}z}\overset{\Delta}{=}\overline{\left(\frac{e^{z}+e^{-z}}{2}\right)} \tag{2.1.41}$$

They are conjugate analytic in all planes and have the following characteristics:

$$\overline{\text{sh}z}=i\ \overline{\sin(iz)},\ \overline{\text{ch}z}=\overline{\cos(iz)} \tag{2.1.42}$$

$$(\overline{\text{ch}z})^2-(\overline{\text{sh}z})^2=1 \tag{2.1.43}$$

$$(\overline{\text{sh}z})^{\circ}=\overline{\text{ch}z},\ (\overline{\text{ch}z})^{\circ}=\overline{\text{sh}z} \tag{2.1.44}$$

1.4.6 Inverse hyperbolic functions

$$\text{Arc}\ \overline{\text{sh}z}=\text{Ln}(\bar{z}+\sqrt{(\bar{z})^2+1}) \tag{2.1.45}$$

$$\text{Arc}\ \overline{\text{ch}z}=\text{Ln}(\bar{z}+\sqrt{(\bar{z})^2-1}) \tag{2.1.46}$$

Chapter Ⅱ Integral Theory of Conjugate Analytic Function

2.1 Conjugate Integral of Complex Function

A new concept—conjugate integral is introduced into this section.

Definition 2.1 Let $C: z=x(t)+iy(t)(\alpha \leqslant t \leqslant \beta)$ be a rectifiable curve in the complex plane, where $z_0[z_0=z(\alpha)]$ is an initial point, $z'[z'=z(\beta)]$ is a terminal point. Let $f(z)=u(x, y)+iv(x, y)$ be continuous on C. Dividing interval $[\alpha, \beta]$ into

$$\alpha=t_0<t_1<\cdots<t_{n-1}<t_n=\beta$$

Thus, there are corresponding points

$$z_0=z(\alpha),\ z_1=z(t_1),\ \cdots,\ z_{n-1}=z(t_{n-1}),\ z_n=z'=z(\beta)$$

on C and they divided C into n sections.

Taking sum

$$\sum_{k=0}^{n-1} f(z'_k)\,\overline{\Delta z_k} \tag{2.2.1}$$

where $\overline{\Delta z_k}=\overline{z_{k+1}}-\overline{z_k}$ and z'_k is an arbitrary point on arc $\widehat{z_k z_{k+1}}$.

Let

$$d=\max \widehat{z_k z_{k+1}} \quad (k=0,1,\cdots,n-1).$$

If

$$\lim_{d \to 0} \sum_{k=0}^{n-1} f(z'_k)\,\overline{\Delta z_k}$$

is existent, no matter how z_k is taken on C and z'_k is taken on $\widehat{z_k z_{k+1}}$, then it is called conjugate integral on oriended curve C, and denoted by

$$\int_C f(z)\,\overline{\mathrm{d}z}=\lim_{d \to 0} \sum_{k=0}^{n-1} f(z'_k)\,\overline{\Delta z_k} \tag{2.2.2}$$

For conjugate integral, we have

Theorem 2.1 Let C be a rectifiable curve in a complex plane, and $f(z)$ be continuous on C.

Then

$$\int_C \alpha f(z)\,\overline{\mathrm{d}z} = \alpha \int_C f(z)\,\overline{\mathrm{d}z}, \tag{2.2.3}$$

where α is a constant.

$$\int_C [f_1(z) + f_2(z)]\,\overline{\mathrm{d}z} = \int_C f_1(z)\,\overline{\mathrm{d}z} + \int_C f_2(z)\,\overline{\mathrm{d}z} \tag{2.2.4}$$

where $f_1(z)$ and $f_2(z)$ are all continuous on C.

$$\int_C f(z)\,\overline{\mathrm{d}z} = \int_{C_1} f(z)\,\overline{\mathrm{d}z} + \int_{C_2} f(z)\,\overline{\mathrm{d}z} \tag{2.2.5}$$

where $C = C_1 + C_2$.

$$\int_C f(z)\,\overline{\mathrm{d}z} = -\int_{C^-} f(z)\,\overline{\mathrm{d}z} \tag{2.2.6}$$

where C^- and C are the same curve, but orientation of C^- is contrary to that of C.

$$\left|\int_C f(z)\,\overline{\mathrm{d}z}\right| \leqslant \int_C |f(z)|\,\mathrm{d}s \leqslant ML \tag{2.2.7}$$

where $|f(z)| \leqslant M$ on C, and L is the arc length of C.

2.2 Integral Theorem

In this section we obtain integral theorem for conjugate analytic function.

Theorem 2.2 Let $f(z)$ be conjugate analytic in domain D, C be an arbitrary rectifiable closed curve in D, and interior of C belongs to D. Then

$$\int_C f(z)\,\overline{\mathrm{d}z} = 0 \tag{2.2.8}$$

Theorem 2.3 Let $f(z)$ be conjugate analytic in a simply connected domain D. If z_0 is a fixed point in D and C is any rectifiable curve in D with initial point z_0 and terminal point z, then the new function is given by

$$F(z) = \int_C f(\xi)\,\overline{\mathrm{d}\xi} = \int_{z_0}^{z} f(\xi)\,\overline{\mathrm{d}\xi} \tag{2.2.9}$$

Theorem 2.4 Let D be interior of rectifiable closed curver C, $f(z)$

be conjugate analytic in D and continuous on $\overline{D}=D+C$, then

$$\int_C f(z)\,\overline{dz}=0 \qquad (2.2.10)$$

Theorem 2.5 Let D be a complex connected domain and $C=C_0+C_1+\cdots+C_n$ be a bound of D (C_i is a contour, $i=0, 1, \cdots, n$). If $f(z)$ is conjugate analytic in D and continuous on $\overline{D}=D+C$, then

$$\int_C f(z)\,\overline{dz}=0 \qquad (2.2.11)$$

Note: Contour is a rectifiable closed curve.

2.3 Oringinal Function and Indefinite Integral

Let D be a domain in a complex plane and $f(z)$ be continuous in D. We have

Theorem 2.6 If $\int_C f(z)\,\overline{dz}=0$ for an arbitrary contour $C\subset D$, then

$$F(z)=\int_{z_0}^{z} f(\xi)\,\overline{d\xi} \qquad (2.2.12)$$

is conjugate analytic in D, and

$$F^{\circ}(z)=f(z)$$

Definition 2.2 Let D be a domain in a complex plane and $f(z)$ be continuous in D. If $\Phi^{\circ}(z)=f(z)$ in D, then $\Phi(z)$ is called an antiderivative of $f(z)$. The set of all antiderivatives of $f(z)$ is called indefinite integral of $f(z)$, and is denoted by

$$\int f(z)\,\overline{dz} \qquad (2.2.13)$$

Theorem 2.7 Let $f(z)$ be continuous in domain D and $\int_C f(z)\,\overline{dz}=0$ for an arbitrary contour $C\subset D$. Then the indefinite integral of $f(z)$ has the following form

$$\Phi(z)=\int_{z_0}^{z} f(\xi)\,\overline{d\xi}+C \quad (z_0\in D) \qquad (2.2.14)$$

where C is an arbitrary constant.

If $\varphi(z)$ is an arbitrary antiderivative of $f(z)$, then

$$\int_{z_0}^{z} f(\xi)\,\overline{d\xi} = \varphi(z) - \varphi(z_0) \tag{2.2.15}$$

Theorem 2.8 Let D be a simply connected domain in a complex plane, and $f(z)$ be continuous in D. If $\int_C f(z)\,\overline{dz} = 0$, for an arbitrary contour $C \subset D$, then the $f(z)$ is conjugate analytic in D.

Theorem 2.9 Let D be a simply connected domain in a complex plane. A $f(z)$ is conjugate analytic in D if and only if $\int_C f(z)\,\overline{dz} = 0$ for an arbitrary contour $C \subset D$.

2.4 Integral Formulas

Let D be a complex connected domain and $C = C_0 + C_1 + \cdots + C_n$ be bound of D (C_i is contour, $i = 0, 1, \cdots, n$). It has

Theorem 2.10 If $f(z)$ is conjugate analytic in D and continuous on $\overline{D} = D + C$, then

$$f(z) = \frac{-1}{2\pi i}\int_C \frac{f(\xi)}{\overline{\xi - z}}\,\overline{d\xi} \tag{2.2.16}$$

for all $z \in D$.

Corollary 2.1 Let $f(z)$ be conjugate analytic in domain $|z - z_0| < R$, and continuous on closed domain $|z - z_0| \leqslant R$. Then

$$f(z_0) = \frac{1}{2\pi}\int_0^{2\pi} f(z_0 + re^{i\theta})\,d\theta \quad (0 < r \leqslant R) \tag{2.2.17}$$

Theorem 2.11 Let D_1 be a domain containing infinite point ∞, and $C = C_1 + C_2 + \cdots + C_n$ be bound of D_1. Let D_1 be always kept on the left of us as we move around C. Let $f(z)$ be conjugate analytic in D_1 (except for ∞), $f(z)$ be continuous on $\overline{D} = D_1 + C$, and $\lim_{z \to \infty} f(z) = f(\infty)$. Then

$$f(z) = \frac{-1}{2\pi i}\int_C \frac{f(\xi)}{\overline{\xi - z}}\,\overline{d\xi} + f(\infty) \tag{2.2.18}$$

for every $z \in D_1$.

Theorem 2.12 Let D be a complex connected domain, and $C = C_0 + C_1 + C_2 + \cdots + C_n$ be bound of D (C_i is contour, $i = 0, 1, \cdots, n$). Let

$f(z)$ be conjugate analytic in D and continuous on $\overline{D}=D+C$. Then $f(z)$ has conjugate derivatives of arbitrary order, and

$$f^{[n]}(z)=\frac{-n!}{2\pi i}\int_C \frac{f(\xi)}{\overline{(\xi-z)^{n+1}}}\overline{\mathrm{d}\xi} \quad (z\in D) \tag{2.2.19}$$

Theorem 2.13 Let $f(z)$ be conjugate analytic and $|f(z)|\leqslant M$ in disc $|z-z_0|<R$. Then

$$|f^{[n]}(z_0)|\leqslant\frac{Mn!}{R^n} \quad (n=1,2,\cdots) \tag{2.2.20}$$

Theorem 2.14 Let $f(z)$ be conjugate analytic and bounded in a complex plane. Then

$$f(z)=\text{constant} \tag{2.2.21}$$

Chapter Ⅲ Series Theory of Conjugate Analytic Function

3.1 Series with Function Term

Theorem 3.1 Let $f_n(z) \in \{f_n(z)\}$ $(n = 1, 2, \cdots)$ be conjugate analytic in domain D, and $\sum_{n=1}^{\infty} f_n(z)$ converge uniformly to $f(z)$ in D. That is,

$$\sum_{n=1}^{\infty} f_n(z) = f(z) \tag{2.3.1}$$

Then $f(z)$ is conjugate analytic in D, and

$$f^{[p]}(z) = \sum_{k=1}^{\infty} f_k^{[p]}(z) \quad (p = 1, 2, \cdots) \tag{2.3.2}$$

where its convergece is uniform on closed $\overline{D}$.

Theorem 3.2 Let D be a domain in complex plane, and ∂D be a bound of D. Let $f_n(z)(n=1,2,\cdots)$ be conjugate analytic in D, $f_n(z)$ $(n=1,2,\cdots)$ be continuous on $\overline{D} = D + \partial D$, and $\sum_{n=1}^{\infty} f_n(z)$ be convergent uniformly on ∂D. Then $\sum_{n=1}^{\infty} f_n(z)$ is convergent uniformly on $\overline{D}$.

3.2 Conjugate Power Series

Definition 3.1 Series of type

$$\sum_{n=0}^{\infty} C_n \overline{(z - z_0)^n} = C_0 + C_1 \overline{(z - z_0)} + C_2 \overline{(z - z_0)^2} + \cdots$$

is called a conjugate power series.

In general, we consider series of type

$$\sum_{n=0}^{\infty} C_n \bar{z}^n = C_0 + C_1 \bar{z} + C_2 \bar{z}^2 + \cdots \tag{2.3.3}$$

Theorem 3.3 If $\sum_{n=0}^{\infty} C_n \bar{z}^n$ is convergent at z_0 ($z_0 \neq 0$), then it is absolute convergent in $|z| < |z_0|$, and convergent uniformly on $|z| \leqslant \rho$ for an arbitrary $\rho \leqslant |z_0|$.

Corollary 3.1 If $\sum_{n=0}^{\infty} C_n \bar{z}^n$ is inconvergent at $z_1 \neq 0$, then it is inconvergent in $|z| > |z_1|$.

Series of type (2.3.3) is always convergent at $z (z=0)$.

Definition 3.2 If $\sum_{n=0}^{\infty} C_n \bar{z}^n$ is convergent in $|z| < R$, and inconvergent in $|z| > R$, then R is called a convergent radius of $\sum_{n=0}^{\infty} C_n \bar{z}^n$, $|z| < R$ and $|z| = R$ are called convergent circular disc and convergence circle of $\sum_{n=0}^{\infty} C_n \bar{z}^n$ respectively.

Theorem 3.4 Let $l = \overline{\lim_{n \to \infty}} \sqrt[n]{|C_n|}$. Then convergence radius of $\sum_{n=0}^{\infty} C_n \overline{(z - z_0)}^n$

$$R = \frac{1}{l} \tag{2.3.4}$$

Theorem 3.5 For $\sum_{n=0}^{\infty} C_n \overline{(z - z_0)}^n$, if $\lim_{n \to \infty} \left| \frac{C_{n+1}}{C_n} \right|$ is existent, then $\lim_{n \to \infty} \sqrt[n]{|C_n|}$ is exsitent, and $\lim_{n \to \infty} \left| \frac{C_{n+1}}{C_n} \right| = \frac{1}{R}$, that is, $R = \lim_{n \to \infty} \left| \frac{C_n}{C_{n+1}} \right|$.

Theorem 3.6 Let $K: |z - z_0| < R (0 < R < +\infty)$ be a convergent disc of $f(z) = \sum_{n=0}^{\infty} C_n \overline{(z - z_0)}^n$. Then

$$f(z) = \sum_{n=0}^{\infty} C_n \overline{(z - z_0)}^n \tag{2.3.5}$$

is conjugate analytic in K.

$$f^{[p]}(z) = p!\ C_p + (p+1)p \cdots 2 C_{p+1} \overline{(z - z_0)} + \cdots + n(n-1) \cdots (n-p+1) C_n \overline{(z - z_0)}^{n-p} + \cdots \tag{2.3.6}$$

in K,

$$C_p=\frac{f^{[p]}(z_0)}{p!} \quad (p=0,1,2,\cdots) \tag{2.3.7}$$

3.3 Series Representations

Theorem 3.7 Let $f(z)$ be conjugate analytic in domain D, and D contain $K:|z-z_0|<R$. Then $f(z)$ has conjugate power series representation.

$$f(z)=\sum_{n=0}^{\infty}C_n\overline{(z-z_0)^n} \quad \text{in} \quad K \tag{2.3.8}$$

where

$$C_n=\frac{f^{[n]}(z_0)}{n!} \quad (n=0,1,2,\cdots) \tag{2.3.9}$$

The above expansion is unique.

Equation (2.3.8) is called a conjugate power series of $f(z)$ at z_0.

Theorem 3.8 A $f(z)$ is conjugate analytic in domain D if and only if $f(z)$ can be expressed in the form

$$f(z)=\sum_{n=0}^{\infty}C_n\overline{(z-z_0)^n} \tag{2.3.10}$$

for every $z_0\in D$.

Theorem 3.9 Let $R>0$ be a convergent radius of $\sum_{n=0}^{\infty}C_n\overline{(z-z_0)^n}$, and $f(z)=\sum_{n=0}^{\infty}C_n\overline{(z-z_0)^n}(z\in k:|z-z_0|<R)$. Then there is at least a singularity on $C:|z-z_0|=R$.

Corollary 3.2 If $f(z)$ is conjugate analytic at z_0, then $f(z)$ can be expressed in the form

$$f(z)=\sum_{n=0}^{\infty}C_n\overline{(z-z_0)^n} \tag{2.3.11}$$

in a neighborhood of z_0 and convergence circle of the series passes through a singularity which is the nearst to z_0 among all singularities of $f(z)$.

3.4 Zeros and Uniqueness Theorem

Definition 3.3 Let $f(z)$ be conjugate analytic in D, and $f(z_0)=0(z_0 \in D)$. Then z_0 is called a zero of $f(z)$.

Theorem 3.10 Let $f(z)$ be conjugate analytic in $|z-z_0|<R$, $f(z)\not\equiv 0$, and $f(z_0)=0$. Then there is a neighborhood of z_0, where there is only z_0 such that $f(z_0)=0$.

Corollary 3.3 Let $f(z)$ be conjugate analytic in $|z-z_0|<R$. If $\{z_n\}\to z_0$ $(z_n \neq z_0)$, where $f(z_n)=0(n=1,2,\cdots)$, then $f(z)\equiv 0$ in $|z-z_0|<R$.

Theorem 3.11 Let $f_1(z)$ and $f_2(z)$ be conjugate analytic in domain D, and $\{z_n\}\to z_0$ $(z_0 \in D, z_n \neq z_0)$. If $f_1(z_n)=f_2(z_n)$ $(n=1,2,\cdots)$, then

$$f_1(z)=f_2(z) \quad \text{for all points in } D. \tag{2.3.12}$$

Corollary 3.4 Let $f_1(z)$ and $f_2(z)$ be conjugate analytic in domain D.

If $f_1(z)=f_2(z)$ in a sub-domain of D, then $f_1(z)=f_2(z)$ in D.

All identical equations, which are established in real case, are established in complex case too. Both sides of the identical equations are conjugate analytic in z-plane.

Theorem 3.12 Let $f(z)$ be conjugate analytic in domain D, and $f(z)\not\equiv$constant. Then there is no maximal value of $|f(z)|$ in D.

Corollary 3.5 Let $f(z)$ be conjugate analytic in domain D and continuous in closed domain $\overline{D}$, and $|f(z)|\leqslant M(z\in\overline{D})$. Then

$$|f(z)|<M(z\in D), \text{except } f(z)=\text{constant}. \tag{2.3.13}$$

Chapter Ⅳ　Isolated Points and Conjugate Power Series with Two-direction

4.1　Conjugate Power Series with Two-direction

Definition 4.1　$\sum\limits_{n=-\infty}^{\infty} C_n \overline{(z-z_0)}^n$ is called conjugate power series with two-direction.

Theorem 4.1　$\sum\limits_{n=-\infty}^{\infty} C_n \overline{(z-z_0)}^n$ is a conjugate analytic function in a ring $K: r<|z-z_0|<R$, and convergent uniformly on closed ring $H: r' \leqslant |z-z_0| \leqslant R' (r<r'<R'<R)$.

Theorem 4.2　Let $f(z)$ be conjugate analytic in ring $H: r<|z-z_0|<R (r \geqslant 0. R<+\infty)$. Then the $f(z)$ can be expressed in the form

$$f(z)=\sum_{n=-\infty}^{\infty} C_n \overline{(z-z_0)}^n \tag{2.4.1}$$

where

$$C_n=\frac{-1}{2\pi i}\int_{\Gamma} \frac{f(\xi)}{\overline{(\xi-z_0)^{n+1}}} \overline{\mathrm{d}\xi} \quad (n=0, \pm 1, \cdots) \tag{2.4.2}$$

and the expansion (2.4.1) is unique.

Definition 4.2　The series (2.4.1) is called an expansion for conjugate power series with two directions of $f(z)$ at z_0, and the series (2.4.2) are called coefficients of expansion (2.4.1).

Definition 4.3　If $f(z)$ is conjugate analytic in a punctured disc $0<|z-z_0|<R$ and is not conjugate analytic at z_0, then z_0 is called an isolated singularity of $f(z)$.

Theorem 4.3　If z_0 is an isolated singularity of $f(z)$, then there is a $R>0$ such that $f(z)=\sum\limits_{n=-\infty}^{\infty} C_n \overline{(z-z_0)}^n$ in punctured disc $0<|z-z_0|<R$.

4.2 Isolated Singularities

Definition 4.4 Let $f(z)=\sum_{n=-\infty}^{\infty} C_n (z-z_0)^n$, where z_0 is an isolated singularity of $f(z)$. Then

(1) $\sum_{n=0}^{\infty} C_n (z-z_0)^n$ is called regular part of $f(z)$ at z_0.

(2) $\sum_{n=1}^{\infty} C_{-n} (z-z_0)^{-n}$ is called principal part of $f(z)$ at z_0.

Definition 4.5 Let z_0 be an isolated singularity.

(1) If the principal part of $f(z)$ at z_0 vanishes, then z_0 is called a removable singularity of $f(z)$.

(2) If the principal part of $f(z)$ has finite terms:

$$\frac{C_{-m}}{(z-z_0)^m}+\frac{C_{-(m-1)}}{(z-z_0)^{m-1}}+\cdots+\frac{C_{-1}}{z-z_0}(C_{-m}\neq 0) \qquad (2.4.3)$$

then z_0 is called a pole of order m of $f(z)$.

(3) If the principal part of $f(z)$ has infinite terms, then z_0 is called an essential singularity of $f(z)$.

For removable singularity, we have

Theorem 4.4 If z_0 is an isolated singularity of $f(z)$, then the following three conditions are equivalent:

(1) The principal part of $f(z)$ at z_0 vanishes.

(2) $\lim_{z\to z_0} f(z)=b \quad (b\neq\infty)$.

(3) $f(z)$ is bounded in a punctured neighborhood $k-\{z_0\}$.

For poles, we have

Theorem 4.5 If z_0 is an isolated singularity, then the following three conditions are equivalent:

(1) The principal part of $f(z)$ at z_0 is

$$\frac{C_{-m}}{(z-z_0)^m}+\cdots+\frac{C_{-1}}{z-z_0} \quad (C_{-m}\neq o)$$

(2) In $k-\{z_0\}$ $f(z)$ can be expressed in the form

$$f(z)=\frac{\lambda(z)}{(z-z_0)^m}$$

where $\lambda(z)$ is conjugate analytic at z_0, and $\lambda(z_0)\neq 0$.

(3) z_0 is a zero of order m of $g(z)=\frac{1}{f(z)}$.

Theorem 4.6 z_0 is a pole of $f(z)$ if and only if

$$\lim_{z\to z_0} f(z)=\infty \tag{2.4.4}$$

For essential singularities, we have

Theorem 4.7 z_0 is an essential singularity of $f(z)$ if and only if

$$\lim_{z\to z_0} f(z) \tag{2.4.5}$$

is not existent.

Theorem 4.8 Let z_0 be an essential singularity of $f(z)$, and there exists a neighborhood of z_0 where $f(z)\neq 0$. Then z_0 is an essential singularity of $\frac{1}{f(z)}$.

Theorem 4.9 Let z_0 is an essential singularity of $f(z)$. Then for an arbitrary constant A (finite or infinite) there is a $\{z_n\}$, where $\{z_n\}\to z_0$, such that

$$\lim_{z_n\to z_0} f(z_n)=A \tag{2.4.6}$$

Theorem 4.10 Let z_0 be an essential singularity of $f(z)$. Then for an arbitrary $A\neq\infty$ (except for $A=A_0$ at most) there is a $\{z_n\}$, where $\{z_n\}\to z_0$, such that $f(z_n)=A(n=1,2,\cdots)$.

4.3 Properties of Conjugate Analytic Function at ∞

Definition 4.6 Let $f(z)$ be conjugate analytic in a neighborhood of ∞: $N-\{\infty\}$. Then the ∞ is called an isolated singularity of $f(z)$.

Definition 4.7 If $z'=0$ is a removable singularity, a pole of order m, and an essential singularity of $\varphi(z')$, then $z=\infty$ is correspondly called a removabe singularity, a pole of order m, and an essential singularity of $f(z)$, where

$$\varphi(z')=f\left(\frac{1}{z'}\right)=f(z) \tag{2.4.7}$$

For point ∞, we have

Theorem 4.11 $z=\infty$ is an isolated singularity of $f(z)$ if and only if one of the following conditions is established:

(1) The principal part of $f(z)$ at point ∞ vanishes.

(2) $\lim\limits_{z\to\infty} f(z)=b \quad (b\neq\infty)$.

(3) $f(z)$ is bounded in $N-\{\infty\}$.

Theorem 4.12 $z=\infty$ is a pole of order m of $f(z)$ if and only if one of the following conditions is established:

(1) The principal part of $f(z)$ at ∞ is

$$b_1\bar{z}+b_2\bar{z}^2+\cdots+b_m\bar{z}^m \quad (b_m\neq 0)$$

(2) In $N-\{\infty\}$, $f(z)$ can be expressed in the form

$$f(z)=\mu(z)\bar{z}^m \tag{2.4.8}$$

where $\mu(z)$ is conjugate analytic in punctured neighborhood $N-\{\infty\}$, and $\mu(\infty)\neq 0$.

(3) $z=\infty$ is a zero of order m of $g(z)=\dfrac{1}{f(z)}$.

Theorem 4.13 $z=\infty$ is a pole of $f(z)$ if and only if

$$\lim_{z\to\infty} f(z)=\infty$$

Theorem 4.14 $z=\infty$ is an essential singularity of $f(z)$ if and only if one of the following conditions is established:

(1) The principal part of $f(z)$ at ∞ has infinite terms which does not vanish.

(2) $\lim\limits_{z\to\infty} f(z)$ is not existent.

Chapter Ⅴ　Residue and Its Applications

5.1　Residue

Definition 5.1　Let z_0 be an isolated singularity of $f(z)$. Then $f(z)$ can be expanded into conjugate power series with two-direction in a neighborhood of z_0.

$$f(z)=\sum_{n=-\infty}^{\infty} C_n \overline{(z-z_0)^n} \quad (0<|z-z_0|<R) \tag{2.5.1}$$

where C_{-1} is called residue of $f(z)$ at z_0, and we use the notation

$$\text{res}[f, z_0]=C_{-1} \tag{2.5.2}$$

Theorem 5.1　Let D be a complex connected domain, C be bound of D, $f(z)$ be conjugate analytic in D (except for z_1, z_2, …, z_n) and continuous on $\overline{D}=D+C$ (except for z_1, z_2, …, z_n). Then

$$\int_C f(z)\,\overline{\mathrm{d}z}=-2\pi i\sum_{k=1}^{n}\text{res}[f, z_k] \tag{2.5.3}$$

Theorem 5.2　Let z_0 be a pole of order n of $f(z)$. Let $f(z)=\dfrac{\varphi(z)}{\overline{(z-z_0)^n}}$ where $\varphi(z)$ is conjugate analytic at z_0 and $\varphi(z_0)\neq 0$, then

$$\text{res}[f, z_0]=\frac{\varphi^{[n-1]}(z_0)}{(n-1)!} \tag{2.5.4}$$

Corollary 5.1　Let z_0 be a pole of order 1 of $f(z)$ and $\varphi(z)=\overline{(z-z_0)}f(z)$. Then

$$\text{res}[f, z_0]=\varphi(z_0) \tag{2.5.5}$$

Corollary 5.2　Let z_0 be a pole of order 2 of $f(z)$ and $\varphi(z)=\overline{(z-z_0)^2}f(z)$, then

$$\text{res}[f, z_0]=\varphi^{\circ}(z_0) \tag{2.5.6}$$

Theorem 5.3 Let z_0 be a pole of order 1 of $f(z)=\frac{\varphi(z)}{\psi(z)}$ [where $\varphi(z)$ and $\psi(z)$ are conjugate at z_0, $\varphi(z_0)\neq 0$, $\psi(z_0)=0$, and $\psi^\circ(z_0)\neq 0$]. Then

$$\operatorname{res}[f, z_0]=\frac{\varphi(z_0)}{\psi^\circ(z_0)} \tag{2.5.7}$$

Definition 5.2 Let $z=\infty$ be an isolated singularity of $f(z)$, and $f(z)$ be conjugate analytic in $N-\{\infty\}: 0\leqslant r<|z|<+\infty$. Then

$$\frac{-1}{2\pi i}\int_{\gamma^-} f(z)\,\overline{\mathrm{d}z} \quad (\gamma: |z|=p>r) \tag{2.5.8}$$

is called residue of $f(z)$ at ∞, and denoted by

$$\operatorname{res}[f, \infty] \tag{2.5.9}$$

where γ^- is a positively oriented contour.

Theorem 5.4 Let only z_1, z_2, $\cdots$, z_n, ∞ be isolated singularities of $f(z)$ on extended plane z. Then

$$\sum_{k=1}^{n}\operatorname{res}[f, z_k]+\operatorname{res}[f, \infty]=0 \tag{2.5.10}$$

5.2 The Argument Principle

Theorem 5.5 Let z_0 be a zero of order m of $f(z)$. Then z_0 is surely a pole of order 1 of $\frac{f^\circ(z)}{f(z)}$, and

$$\operatorname{res}\left[\frac{f^\circ(z)}{f(z)}, z_0\right]=m \tag{2.5.11}$$

Let z_1 be a pole of order n of $f(z)$. Then z_1 is a pole of order 1 of $\frac{f^\circ(z)}{f(z)}$, and

$$\operatorname{res}\left[\frac{f^\circ(z)}{f(z)}, z_1\right]=-n \tag{2.5.12}$$

Theorem 5.6 Let C be a contour and $f(z)$ satisfy the following conditions:

(1) $f(z)$ is conjugate analytic in an interior of C (except for a finite number of poles)

(2) $f(z)$ is conjugate analytic and $f(z)\neq 0$ on C.

Then

$$\frac{-1}{2\pi i}\int_C \frac{f^{\circ}(z)}{f(z)}\overline{dz} = N(f,c) - P(f,c) \qquad (2.5.13)$$

where $N(f, c)$ and $P(f, c)$ are number of zeros and poles in an interior of C, respectively.

Theorem 5.7 If $f(z)$ satisfies conditions of the theorem 5.6, then

$$P(f,c)-N(f,c)=\frac{\Delta_C \arg f(z)}{2\pi} \qquad (2.5.14)$$

where Δ_C arg $f(z)$ is an increment of arg $f(z)$ as z moves around once on positive direction.

Theorem 5.8 Let C be a contour, $f(z)$ and $\varphi(z)$ satisfy the following conditions:

(1) They are conjugate analytic on C and in its interior.

(2) $|f(z)|>|\varphi(z)|$ on C.

Then $f(z)$ and $f(z)+\varphi(z)$ have the same number of zeros:

$$N(f+\varphi,c)=N(f,c) \qquad (2.5.15)$$

Theorem 5.9 If $f(z)$ is single-valued conjugate analytic in domain D, then $f^{\circ}(z)\neq 0$ in D.

Chapter Ⅵ Conjugate Analytic Extensions

6.1 Power Series Extension

Definition 6.1 Let $f(z)$ be conjugate analytic in domain D. We consider a greater domain $G \supset D$. If there is $F(z)$ such that $F(z)$ is conjugate analytic in G, and $F(z)=f(z)$ in D, then $F(z)$ is called a conjugate analytic extension of $f(z)$ in G.

Definition 6.2 $\{D, f(z)\}$ is called a conjugate analytic element, where D is a domain and $f(z)$ is conjugate analytic in D. $\{D_1, f_1(z)\}=\{D_2, f_2(z)\}$ if and only if $D_1=D_2$ and $f_1(z)=f_2(z)$.

Theorem 6.1 Let $\{D_1, f_1(z)\}$ and $\{D_2, f_2(z)\}$ be two conjugate analytic elements. If

(1) $D_1 \cap D_2 = d_{12}$, $D_1 \supset D_2$ and $D_2 \supset D_1$,

(2) $f_1(z)=f_2(z)$, $z \in d_{12}$,

then $\{D_1+D_2, F(z)\}$ is also a conjugate analytic element, where

$$F(z)=\begin{cases} f_1(z), & z \in D_1-d_{12} \\ f_1(z)=f_2(z), & z \in d_{12} \\ f_2(z), & z \in D_2-d_{12} \end{cases} \tag{2.6.1}$$

Definition 6.3 Let $D_1 \cap D_2 = d_{12}$ be a domain, $D_1 \not\supset D_2$, $D_2 \not\supset D_1$, and $f_1(z)=f_2(z)$, $z \in d_{12}$. Then $\{D_1, f_1(z)\}$ and $\{D_2, f_2(z)\}$ are mutually called direct conjugate analytic extension.

We consider $\{D_1, f_1(z)\}$. Let z_1 be an arbitrary point in D. Then $f_1(z)$ can be expanded into conjugate power series in a neighborhood of z_1.

$$f_1(z)=\sum_{n=0}^{\infty} C_n^{(1)} \overline{(z-z_1)^n} \tag{2.6.2}$$

Let $R_1>0$ be a conjugate radius of equation (2.6.2). We take $z_2 \in \Gamma_1: |z-z_1|<R_1$, $z_2 \neq z_1$ and expand $f_1(z)$ into conjugate power series in

a neighborhood of z_2.

$$\sum_{n=0}^{\infty} C_n^{(2)} \overline{(z - z_2)^n} \qquad (2.6.3)$$

In genaral using this method $f_1(z)$ may be always extented into a greater domain.

6.2 Lens Arc Extension and Symmetry Principle

Theorem 6.2 Let $\{D_1, f_1(z)\}$ and $\{D_2, f_2(z)\}$ be two conjugate analytic elements.

If

(1) $D_1 \cap D_2 = \Phi$ and open arc Γ is common bound of D_1 and D_2.

(2) $f_1(z)$ is continuous in $D_1 + \Gamma$, $f_2(z)$ is continuous in $D_2 + \Gamma$.

(3) $f_1(z) = f_2(z) \quad z \in \Gamma$.

Then $\{D_1 + \Gamma + D_2, F(z)\}$ is also a conjugate analytic element, where $F(z)$ satisfies the equation (1.2.5).

Definition 6.4 Let $\{D_1, f_1(z)\}$ and $\{D_2, f_2(z)\}$ satisfy conditions of theorem 6.2. Then $\{D_1, f_1(z)\}$ and $\{D_2, f_2(z)\}$ are multually called a lens arc extension.

Theorem 6.3 Let D and D^* be two domains in z-plane, they locate respectively in upper semi-plane and below semi-plane and are symmetric with regard to X-axis, and argment $S = D_1 \cap D_2$ belongs to X-axis. Let $\{D, f(z)\}$ be conjugate analytic element, $f(z)$ be continuous on $D + S$ and real number on S. Then $\{D + S + D^*, F(z)\}$ is also a conjugate analytic element, where $F(z)$ satisfies the equation (1.2.6).

Chapter Ⅶ Anti-conformal Mapping

7.1 Character of Conjugate Analytic Translation

Let C be a smooth arc, represented by the equation $z=z(t)(t_0\leqslant t\leqslant t_1)$, and let $f(z)$ be a function defined at all points on C. The equation $\mathrm{w}=f[z(t)](t_0\leqslant t\leqslant t_1)$ is a parametric representation of the image Γ of C under the transformation $\mathrm{w}=f(z)$.

Suppose that C passes through a point $z_0=z(t_0)$, where $f(z)$ is conjugate analytic and that $f^{\circ}(z_0)\neq 0$

Let $\alpha=\arg f^{\circ}(z_0)$. Then $\alpha=\psi+\varphi$. where φ is angle of inclination of a directed line tangent to C at z_0 and ψ is angle of inclination of a directed line tangent to Γ at $f(z_0)$.

Now let C_1 and C_2 be two smooth arcs passing through z_0. Let φ_1 and φ_2 be angles of inclination of directed lines tangent to C_1 and C_2, respectively, at z_0 and let ψ_1 and ψ_2 be angles of inclination of directed lines tangent to Γ_1 and Γ_2, respectively, at $f(z_0)$.

Then

$$\psi_1-\psi_2=-(\varphi_1-\varphi_2) \tag{2.7.1}$$

So, there are

Theorem 7.1 If $f(z)$ is conjugate analytic at z_0 and $f^{\circ}(z_0)\neq 0$, then transformation $\mathrm{w}=f(z)$ is anti-conformal at point z_0.

Theorem 7.2 Let $\mathrm{w}=f(z)$ be conjugate analytic in domain D. If $f^{\circ}(z)\neq 0$ at $z\in D$, then $f(z)$ is anti-conformal at z.

Corollary 7.1 If $\mathrm{w}=f(z)$ is simple conjugate analytic in D, then $\mathrm{w}=f(z)$ is anti-conformal in D.

Theorem 7.3 Let $\mathrm{w}=f(z)$ be conjugate analytic in domain D and $f(z)\not\equiv$constant. Then $G=f(D)$ is also a domain.

Corollary 7.2 Let $w=f(z)$ be simple conjugate analytic in domain D. Then $G=f(D)$ is also a domain.

Theorem 7.4 Let $w = f(z)$ be simple conjugate analytic in domain D. Then

(1) $w=f(z)$ transfortes D into domain $G=f(D)$ anti-conformally.

(2) Inverse function $z = f^{-1}(w)$ is simple conjugate analytic in D, and

$$[f^{-1}(w_0)]^{\circ}=\frac{1}{f^{\circ}(z_0)} \quad [z_0\in D, w_0=f(z_0)\in G] \qquad (2.7.2)$$

7.2 Linear Transformations

Let a, b, c and d denote four complex constants with the restriction that $ad\neq bc$. Then the transformation $w=\dfrac{a\,\bar{z}+b}{c\,\bar{z}+d}$ is called a linear transformation.

We easily observe that this transformation is a composition of the following simple transformations:

(1) $$w=k\,\bar{z}+h \quad (k\neq 0)$$

(2) $$w=\frac{1}{z}$$

(3) $$w=kz+h$$

For linear transformation, there are

Theorem 7.5 A linear transformation maps the class of circles and lines onto itself.

Theorem 7.6 A linear transformation is anti-conformal in extended plane.

Theorem 7.7 A linear transformation maps four distinct points z_1, z_2, z_3 and z_4 onto four distinct points w_1, w_2, w_3 and w_4, respectively, and

$$\frac{z_4-z_1}{z_4-z_2}:\frac{z_3-z_1}{z_3-z_2}=\overline{\left(\frac{w_4-w_1}{w_4-w_2}\right)}:\overline{\left(\frac{w_3-w_1}{w_3-w_2}\right)} \qquad (2.7.3)$$

Theorem 7.8 There exists a unique linear transformation that maps distinct points z_1, z_2 and z_3 onto three distinct points w_1, w_2 and w_3, respectively. An implicit formula for the mapping is given by the equation

$$\frac{w-w_1}{w-w_2}:\frac{w_3-w_1}{w_3-w_2}=\overline{\left(\frac{z-z_1}{z-z_2}\right)}:\overline{\left(\frac{z_3-z_1}{z_3-z_2}\right)} \tag{2.7.4}$$

Theorem 7.9 Let z_1 and z_2 be symmetric with regard to circle C. Let $w=L(\overline{z})$ be a linear transformation. Then $w_1=L(\overline{z_1})$ and $w_2=L(\overline{z_2})$ are symmetric with regard to circle $\Gamma=L(c)$.

7.3 Some Element Functions

Let α be a real constant. Then the transformation $w=\overline{z^\alpha}=\overline{e^{\alpha \mathrm{Ln} z}}$ is called a conjugate power function.

When $\alpha=0$, w is anti-conformal in complex plane except for 0 and ∞.

When $\alpha>1$, principal value of w maps angle-domain: $-\frac{\pi}{\alpha}<\arg z<\frac{\pi}{\alpha}$ onto the w-plane slit along negative real axis.

When $\alpha<1$, w can't map angle-domain onto a slit w-plane.

The transformation $w=\overline{e^z}$ is called a conjugate exponential function.

It is anti-conformal in z-plane because $(\overline{e^z})^\circ=\overline{e^z}\neq 0$ for an arbitrary point z.

It maps strip-domain: $0<\mathrm{Im} z<h$ $(0\leqslant h\leqslant 2\pi)$ in z-plane onto angle-domain: $-h<\arg w<0$ in w-plane anti-conformally.

The transformation $w=\overline{\mathrm{Ln} z}$ is called a conjugate logarithm function.

It maps angle-domain g: $-h<\arg z<0$ $(0\leqslant h\leqslant 2\pi)$ in z-plane onto strip-domain G: $0<\mathrm{Im}\ w<h$ in w-plane, where $\overline{\mathrm{Ln} z}$ is a conjugate analytic branch in G.

7.4 Existence Theorem and Bounded Correspondence Theorem

Theorem 7.10 Let D be a simply connected domain in extended z-

plane and its bound points be more than one. Then there is only one simple conjugate analytic function $w=f(z)$ in D, and it maps D onto unit circle $|w|<1$, and

$$f(z_0)=0, \quad f^{\circ}(z_0)>0 \quad (z_0 \in D) \tag{2.7.5}$$

Theorem 7.11 Let D and G be simply connected domains, and simple closed curves C and Γ be their bounds, respectively. Let $w=f(z)$ map D onto G anti-conformally. Then there is $F(z)=f(z)$ in D and it is continuous on $\overline{D}=D+C$ and it maps C onto Γ, and the map is bilateral simple and continuous.

Inverse theorem of the above conclution is established too.

Chapter Ⅷ Brief Introduction of Applications

In this chapter we use conjugate analytic function to solve some of physical and geometric problems. Problems in hydrodynamics, electrostatic potential, elastrodynamics, and geometric transformation will be treated.

8.1 Two-dimensional Fluid Flow

We consider only the two-dimensional steadystate type of problem. That is, the motion of fluid is assumed to be the same in all planes parallel to the xy plane, the velocity being parallel to that plane and independent of time.

Let the flow be steady in domain D of z-plane, $v=p+\mathrm{i}q$ be velocity of fluid at $z\in D$, p, q and their partial derivatives p_x, p_y, q_x and q_y be continuous in D.

Thus, we have

Theorem 8.1 A steady flow is harmonic in D if and only if $v=p+\mathrm{i}q$ is conjugate analytic in D.

Since

$$\mathrm{div}\, v=p_x+\mathrm{i}q_y=0$$

there is a $\psi(x, y)$ such that

$$\mathrm{d}\psi(x,y)=q\,\mathrm{d}x-p\,\mathrm{d}y$$

Thus

$$\psi_x=q, \psi_y=-p \tag{2.8.1}$$

So

$$\frac{\mathrm{d}y}{\mathrm{d}x}=-\frac{\psi_x}{\psi_y}=\frac{q}{p} \tag{2.8.2}$$

It is interpreted that $\psi(x,y)=C_1$ is a streamline of the flow (C_1 is a real constant).

Since

$$\text{rot } v = q_x - p_y = 0$$

there is a $\varphi(x, y)$ such that

$$d\varphi(x, y) = p\,dx + q\,dy$$

Thus

$$\varphi_x = p, \quad \varphi_y = q \tag{2.8.3}$$

That is

$$\text{grad}\varphi = (p, q) = v \tag{2.8.4}$$

It is interpreted that $\varphi(x, y)$ is a potential function of the flow.

From (2.8.1) and (2.8.3), we obtain

$$\varphi_x = -\psi_y, \quad \varphi_y = \psi_x \tag{2.8.5}$$

It follows that $f(z) = \varphi(x, y) + i\psi(x, y)$ is conjugate analytic in D and called a complex of the flow in D.

For the $f(z)$, we have

$$f^{\circ}(z) = \varphi_x + i\psi_x = p + iq \tag{2.8.6}$$

It follows that if a flow is harnomic, then both complex and velocity of the flow are conjugate analytic in D, and conjugate derivetive of complex of the flow is velocity of the flow.

Thus, it is quite convenient that we use conjugate analytic function to discribe and study indivergent and irrotational flow. Moreover, we have

$$I = \int_C f^{\circ}(z)\,\overline{dz} = \int_C p\,dx + q\,dy + i\int_C q\,dx - p\,dy = \Gamma + iN \tag{2.8.7}$$

where C is a contour in D, Γ and N are circulation and flow, respectively.

Example 8.1 Let $f(z) = \alpha\bar{z}$ ($\alpha > 0$) be complex of two-dimensional flow. Obtain velocity, stream function, streamline, potential function, equipotential line for the flow, respectively.

Solution.

Velocity $$f^{\circ}(z) = a$$

From

$$f(z) = a\bar{z} = ax - iay \tag{2.8.8}$$

we have

stream function $$-ay$$

streamlines $ay=C_1$

potential function ax

equipotential lines $ax=C_2$

Example 8.2 Let $f(z)=\frac{1}{\bar{z}}$ be a complex of a two-dimensional flow. Obtain velocity, stream function, streamline, potential function, equipotential line of the follow.

Solution.

Velocity $f^{\circ}(z)=-\frac{1}{\bar{z}^2}$

From

$$f(z)=\frac{1}{\bar{z}}=\frac{z}{\bar{z}z}=\frac{x+iy}{x^2+y^2} \tag{2.8.9}$$

we obtain

stream function $\frac{y}{x^2+y^2}$

streamlines $\frac{y}{x^2+y^2}=C_1$

potential function $\frac{x}{x^2+y^2}$

equipotential lines $\frac{x}{x^2+y^2}=C_2$

8.2 Two-dimensional Electrostatics

In two-dimensional electrostatics, electric potential $u(x, y)$ and electric flux $v(x, y)$ are all harmonic functions, and equipotential lines $u(x, y)=C_1$ and electrolines $v(x, y)=C_2$ are orthogonal. The properties are conform to that of real and imagate parts of a conjugate analytic function.

A conjugate analytic function $f(z)=u(x,y)+iv(x,y)$ is called a complex of a electrostatic field if $u(x, y)$ and $v(x, y)$ are electric potential and electric flux of the field, respectively. Moreover, we can see that

$$f^{\circ}(z)=u_x+\mathrm{i}u_y=\mathrm{grad}u$$
$$=-\mathbf{E}(negative\ fieldintensity)$$

That is, a conjugate derivate of complex of electrostic field is a negative field intensity of that.

So, it is faily convenient and extremely visual that we use conjugate analytic function to discribe two-dimensional electrostatics.

Similar to discuss of hydrodynamic, we have

Theorem 8.2 A electrostatic field is tubular and potential if and only if the field intensity is conjugate analytic.

Example 8.3 Let $f(z)=\frac{1}{\bar{z}}$ be a complex of an electrostatic field, obtain electric potential, equipotential lines, electric flux, electrolines and field intensity of the field, respectively.

Solution.

Electric potential $\qquad \frac{x}{x^2+y^2}$

equipotential lines $\qquad \frac{x}{x^2+y^2}=C_1$

Electric flux $\qquad v(x,y)=\frac{y}{x^2+y^2}$

Electrolines $\qquad \frac{y}{x^2+y^2}=C_2$

field intensity $\qquad \mathbf{E}=-f^{\circ}(z)=\frac{1}{\bar{z}^2}$

that is

$$|\mathbf{E}|=\frac{1}{|z|^2},\quad \arg\mathbf{E}=2\arg z$$

8.3 Two-dimensional Elastrodynamics

Using conjugate analytic functions to express stress function, displacement and stress of plane-problem.

For body force is selfweight, soluting plane-problem sum up obtaining stress function φ which satisfies the equation

$$\nabla^2(\nabla^2\varphi)=\nabla^4\varphi=0 \tag{2.8.10}$$

and its bounded conditions.

We can obtain solution of the equation by means of conjugate analytic function, that is, we can use conjugate analytic function to express two-dimensional stress function.

Let

$$P=\nabla^2\varphi \tag{2.8.11}$$

From (2.8.10) we see that $\nabla^2 P=0$, and P is a harmonic function. Take harmonic function Q such that

$$f\ (z)\ =P-\mathrm{i}Q$$

is a conjugate analytic function.

Let

$$\psi(z)=\frac{1}{4}\int f(z)\ \overline{\mathrm{d}z}=p-\mathrm{i}q$$

Then $\psi(z)$ is also conjugate analytic, and

$$\psi^{\circ}(z)=\frac{1}{4}f(z)=\frac{1}{4}(P-\mathrm{i}P)$$

So

$$\frac{\partial p}{\partial x}=\frac{1}{4}P,\quad \frac{\partial q}{\partial y}=\frac{1}{4}P$$

Since p and q are harmonic, $\nabla^2 p=0$ and $\nabla^2 q=0$.

Moreover

$$\nabla^2(xp)=x\ \nabla^2 p+2\frac{\partial p}{\partial x}=2\frac{\partial p}{\partial x}$$

$$\nabla^2(yq)=y\ \nabla^2 q+2\frac{\partial q}{\partial y}=2\frac{\partial q}{\partial y}$$

Hence

$$\nabla^2(xp+yq)=2\frac{\partial p}{\partial x}+2\frac{\partial q}{\partial y}=P$$

$$\nabla^2(\varphi-xp-yq)=\nabla^2\varphi-\nabla^2(xp+yq)=P-P=0$$

Let

$$p_1=\varphi-xp-yq$$

It is obvious that p_1 is a harmonic function.

Then

$$\varphi=xp+yq+p_1 \tag{2.8.12}$$

Introduting harmonic function q_1 such that

$\chi(z)=p_1-\mathrm{i}q_1$ is conjugate analytic.

So

$$z\psi(z)+\chi(z)=(x+\mathrm{i}y)(p-\mathrm{i}q)+p_1-\mathrm{i}q_1$$
$$=(xp+yq+p_1)-\mathrm{i}(xq-py+q_1) \tag{2.8.13}$$

By (2.8.12) and (2.8.13), we obtain

$$\varphi=\mathrm{Re}\big(z\varphi(z)+\chi(z)\big) \tag{2.8.14}$$

or

$$\varphi=\frac{1}{2}[z\psi(z)+\chi(z)+\bar{z}\,\overline{\psi(z)}+\overline{\chi(z)}] \tag{2.8.15}$$

(2.8.14) and (2.8.15) illustrate that an arbitrary two-dimensional stress function can be expressed by a conjugate analytic function, which is choosed suitably.

In the following we can see that two-dimensional displacements and stresses are also expressed by conjugate analytic functions.

For two-dimensional stress problems of no-body force, we have

$$\begin{cases}\dfrac{\partial u}{\partial x}=\dfrac{1}{E}(\sigma_x-v\sigma_y)=\dfrac{1}{E}\left(\dfrac{\partial^2\varphi}{\partial y^2}-v\,\dfrac{\partial^2\varphi}{\partial x^2}\right)\\ \dfrac{\partial v}{\partial y}=\dfrac{1}{E}(\sigma_y-v\sigma_x)=\dfrac{1}{E}\left(\dfrac{\partial^2\varphi}{\partial x^2}-v\,\dfrac{\partial^2\varphi}{\partial y^2}\right)\\ \dfrac{\partial v}{\partial x}+\dfrac{\partial u}{\partial y}=\dfrac{2(1+v)}{E}\tau_{xy}=-\dfrac{2(1+v)}{E}\dfrac{\partial^2\varphi}{\partial x\partial y}\end{cases} \tag{2.8.16}$$

where u and v are displacements, and φ is a stress function.

By

$$P=\nabla^2\varphi=\frac{\partial^2\varphi}{\partial x^2}+\frac{\partial^2\varphi}{\partial y^2},\quad \frac{\partial p}{\partial x}=\frac{\partial q}{\partial y}=\frac{1}{4}P$$

the first and second formulas of (2.8.16) translate into

$$\begin{cases}\dfrac{\partial u}{\partial x}=\dfrac{1}{E}\left[4\dfrac{\partial p}{\partial x}-(1+v)\dfrac{\partial^2 \varphi}{\partial x^2}\right]\\ \dfrac{\partial v}{\partial y}=\dfrac{1}{E}\left[4\dfrac{\partial q}{\partial y}-(1+v)\dfrac{\partial^2 \varphi}{\partial y^2}\right]\end{cases} \tag{2.8.17}$$

Integrating two sides of (2.8.17), we decude

$$\begin{cases}u=\dfrac{1}{E}\left[4p-(1+v)\dfrac{\partial \varphi}{\partial x}+g_1(y)\right]\\ v=\dfrac{1}{E}\left[4q-(1+v)\dfrac{\partial \varphi}{\partial y}+g_2(x)\right]\end{cases} \tag{2.8.18}$$

By (2.8.18), the third formula of (2.8.16) translates into

$$4\left(\frac{\partial p}{\partial y}+\frac{\partial q}{\partial x}\right)-2(1+v)\frac{\partial^2 \varphi}{\partial x \partial y}+\frac{\mathrm{d}g_1}{\mathrm{d}y}+\frac{\mathrm{d}g_2}{\mathrm{d}x}=-2(1+v)\frac{\partial^2 \varphi}{\partial x \partial y}$$

By

$$\frac{\partial p}{\partial y}+\frac{\partial q}{\partial x}=0$$

we obtain

$$\frac{\mathrm{d}g_1}{\mathrm{d}y}+\frac{\mathrm{d}g_2}{\mathrm{d}x}=0$$

So

$$\frac{\mathrm{d}g_1}{\mathrm{d}y}=-\frac{\mathrm{d}g_2}{\mathrm{d}x}=C,(C \text{ is constant})$$

$$g_1(y)=Cy+C_1,\ g_2(x)=-Cx+C_2$$

Eliminate $g_1(y)$ and $g_2(x)$, then (2.8.18) translates into

$$\begin{cases}u=\dfrac{1}{E}\left[4p-(1+v)\dfrac{\partial \varphi}{\partial x}\right]\\ v=\dfrac{1}{E}\left[4q-(1+v)\dfrac{\partial \varphi}{\partial y}\right]\end{cases} \tag{2.8.19}$$

or

$$u+\mathrm{i}v=\frac{1}{E}\left[4(p+\mathrm{i}q)-(1+v)\left(\frac{\partial \varphi}{\partial x}+\mathrm{i}\frac{\partial \varphi}{\partial y}\right)\right] \tag{2.8.20}$$

By (2.8.15), we deduce

$$\frac{\partial \varphi}{\partial x}+\mathrm{i}\frac{\partial \varphi}{\partial y}=\overline{\psi(z)+z\psi^{\circ}(z)+\chi^{\circ}(z)} \tag{2.8.21}$$

Substituting in (2. 8. 20) by (2. 8. 21) and noting that $p + \mathrm{i}q = \psi(z)$, we find

$$u+\mathrm{i}v=\frac{3-v}{E}\overline{\psi(z)}+\frac{1+v}{E}[z\psi^{\circ}(z)+\chi^{\circ}(z)] \qquad (2.8.22)$$

The (2. 8. 22) is an expression of u and v by conjugate analytic functions $\psi(z)$ and $\chi(z)$.

To obtain σ_x, σ_y and τ_{xy}, we deferente two sides of (2. 8. 21) to x

$$\frac{\partial^2\varphi}{\partial x^2}+\mathrm{i}\frac{\partial^2\varphi}{\partial x\partial y}=\overline{\psi^{\circ}(z)}+z\psi^{[2]}(z)+\psi^{\circ}(z)+\chi^{[2]}(z) \qquad (2.8.23)$$

and deferente two sides of (2. 8. 21) to y and multiply that of (2. 8. 21) by i

$$\mathrm{i}\frac{\partial^2\varphi}{\partial x\partial y}-\frac{\partial^2\varphi}{\partial y^2}=-\overline{\psi^{\circ}(z)}+z\psi^{[2]}(z)-\psi^{\circ}(z)-\chi^{[2]}(z) \qquad (2.8.24)$$

Subtract (2. 8. 24) from (2. 8. 23)

$$\sigma_x+\sigma_y=2\overline{\psi^{\circ}(z)}+2\psi^{\circ}(z)=4\mathrm{Re}\psi^{\circ}(z) \qquad (2.8.25)$$

Add (2. 8. 23) to (2. 8. 24)

$$\sigma_y-\sigma_x-2\mathrm{i}\tau_{xy}=2[z\psi^{[2]}(z)+\chi^{[2]}(z)] \qquad (2.8.26)$$

Both (2. 8. 25) and (2. 8. 26) illustrate that σ_x, σ_y and τ_{xy} can all be expressed by conjugate analytic functions $\psi(z)$ and $\chi(z)$.

8. 4 Applications of Anti-conformal Translation

The conjugate analytic function has a property of anti-conformal translation, so we may use this property to obtain complex of a plane-field under given bounded conditions. That is, translating a plane-field which has a more complex boundary form into a plane-field which has a more simple boundary that by means of anti-conformal translation.

In the following we will illustrate with examples.

Example 8. 4 A is a very large conductor of metal which was diged 60°-angle of twosides. Let's charge the A until it has eletric potential V_0, obtain electric potential distribution in the angle of two-sides.

Solution. Consider the conductor as infinite long. It is only need to study a section of A and call the section z-plane. In the z-plane the 60°-angle region expresses the angle of twosides (Figure 1).

Enlarge the 60°-angle to lower half-plane (Figure 2).

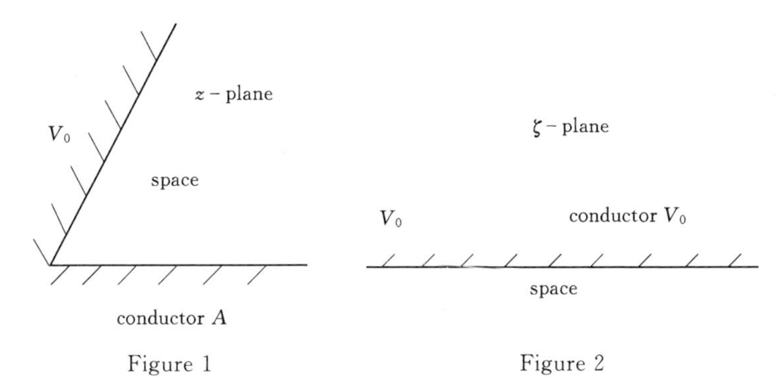

Figure 1

Figure 2

In fact, the transformation $\zeta=\overline{z^3}$ can finish the above enlarge. In ζ-plane upper half-plane is the conductor, and lower half-plane is the space.

A electric potential distribution of lower half-plane is very obvious.

$$u=V_0-C\eta \quad (\zeta=\xi+\mathrm{i}\eta) \tag{2.8.27}$$

where C is dependent on a charge density of the conductor face.

Returning to z-plane, the electric potential distribution in the angle region

$$\begin{aligned} u &=V_0-C\mathrm{Im}\zeta=V_0-C\mathrm{Im}\,\overline{z^3} \\ &=V_0+C(3x^2y-y^3) \end{aligned} \tag{2.8.28}$$

Example 8.5 Figure 3 expresses a flow in a water trough with plane-botton. There is a vertical slice on the trough. Obtain velocity and complex of the flow.

Solution. A deficulty of this problem comes from the slice on the trough bottom, there is a right angle on two sides of the slice, respectively.

The function

$$z_1=z^2 \tag{2.8.29}$$

maps upper half-plane with slice onto z_1-plane slit along ray $y_1=0$, $x_1\geqslant-h^2$ (Figure 4).

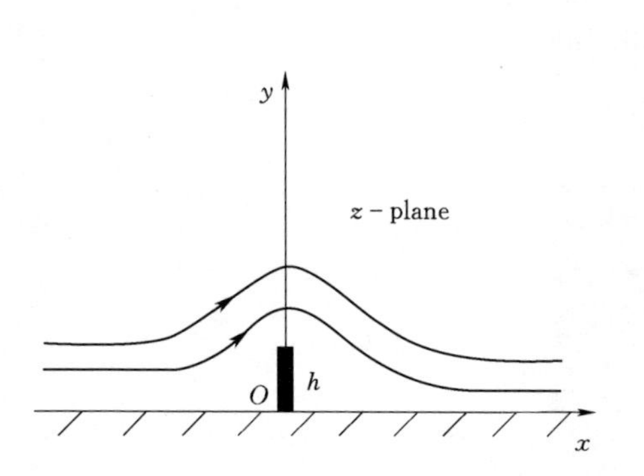

Figure 3

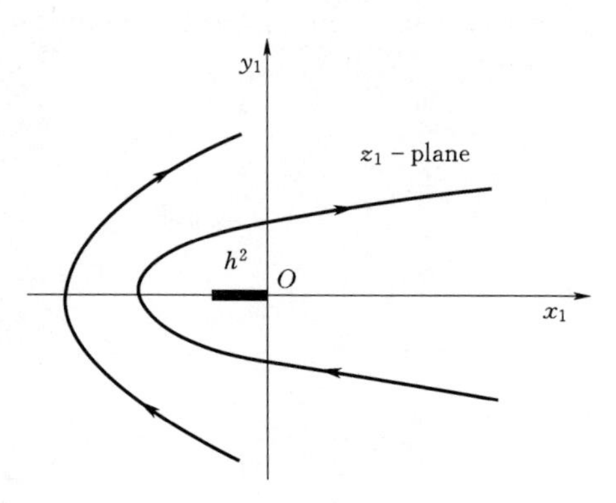

Figure 4

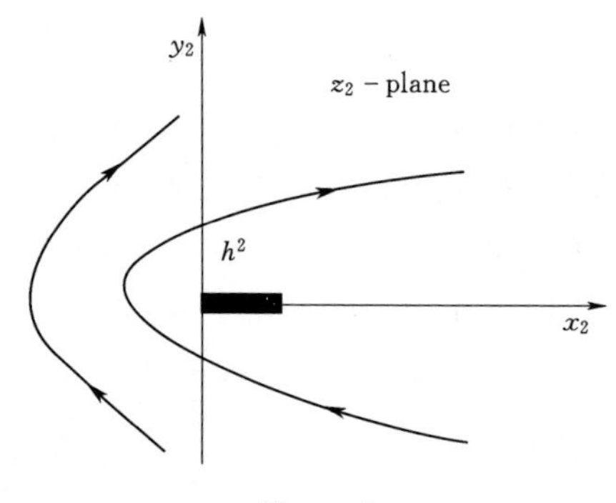

Figure 5

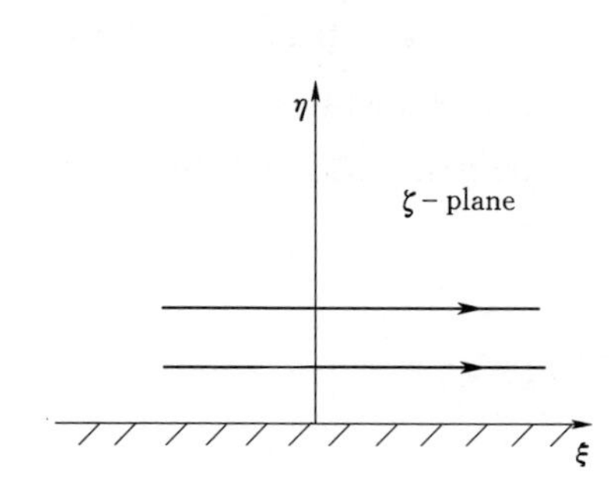

Figure 6

The function

$$z_2 = z_1 + h^2 \quad (2.8.30)$$

maps endpoint of the slit onto origin in z_2-plane (Figure 5).

The function

$$\zeta = \sqrt{z_2} \quad (2.8.31)$$

maps z_2-plane onto the upper half-plane of ζ-plane, and maps the slit onto real axis of ζ-plane (Figure 6).

In ζ-plane the complex of the flow is

$$f(\zeta) = C\bar{\zeta} \quad (C > 0) \quad (2.8.32)$$

Returning to z-plane

$$C\bar{\zeta} = C\overline{\sqrt{z_2}} = C\overline{\sqrt{z_1 + h^2}} = C\overline{\sqrt{z^2 + h^2}} \quad (2.8.33)$$

This is complex of orginal flow.

The velocity of the flow

$$\left(C\sqrt{z^2+h^2}\right)^{\circ}=C\overline{\left(\frac{z}{\sqrt{z^2+h^2}}\right)} \tag{2.8.34}$$

Since

$$\lim_{z\to\infty} C\overline{\left(\frac{z}{\sqrt{z^2+h^2}}\right)}=C\overline{\left(\lim_{z\to\infty}\frac{z}{\sqrt{z^2+h^2}}\right)}=C \tag{2.8.35}$$

C is a velocity of the flow at the point far from the slice.

PART Ⅲ Their Tremendous Influences

Chapter Ⅰ Some Properties of Bianalytic Function

Due to their wonderful properties, analytic functions possess a lot of application in mechanics and physics-mathematics. For example, when we research the physical fields on plane without source and curl, theories of analytic function display their strength forces. But if discuss the phsical fields with sources or curls, then this important tool of analytic function will has no use.

As everyone knows, the complex form of C-R equation is

$$\frac{\partial \mathrm{w}}{\partial \bar{z}}=0 \tag{3.1.1}$$

here $\frac{\partial}{\partial \bar{z}}=\frac{1}{2}\left(\frac{\partial}{\partial x}+\mathrm{i}\frac{\partial}{\partial y}\right), \mathrm{w}=u+\mathrm{i}v, z=x+\mathrm{i}y$

Suppose G is a rigion on plant and define in G a complex function $\mathrm{w}(z)$, which exists the two order derivitive $\frac{\partial^2 \mathrm{w}}{\partial \bar{z}^2}$ on $\bar{z}$.

If $\mathrm{w}(z)$ sufisfies the following partial differential equation

$$\overline{\frac{\partial \mathrm{w}}{\partial \bar{z}}}=f$$

here, f is an analytic function, then

$$\frac{\partial^2 \mathrm{w}}{\partial \bar{z}^2}=0 \quad z\in G \tag{3.1.2}$$

and we will say $\mathrm{w}(z)$ is a bianalytic function in G.

obviously, $$\mathrm{w}(z)=\frac{-1}{\pi}\iint_G \frac{f(\xi,\eta)}{\zeta-z}\mathrm{d}\xi\mathrm{d}\eta+\Phi(z) \tag{3.1.3}$$

where $\zeta=\xi+\mathrm{i}\eta\in G$, and $\Phi(z)$ is an analytic function.

1.1 Bianalytic Functions Uniqueness

Let's concider the following boundary value question:

Question A

$$\left.\frac{\partial w}{\partial \bar{z}}\right|_{\Gamma}=0 \tag{3.1.4}$$

$$w|_{\Gamma}=0 \tag{3.1.5}$$

we will seek a $w(z)$ that satisfies (3.1.2) here $w(z)$ is regular and Γ is boundary of G.

Theorem 1.1 Question A is only $w(z)\equiv 0$, $z\in G$.

In fact, from (3.1.2) will have $w(z)$ is analytic function in G, further from (3.1.4) will have $\frac{\partial w}{\partial \bar{z}}=0$, $z\in G$. then $w(z)$ is an analytic function in G and from (3.1.5) $w|_{\Gamma}=0$ will be $w(z)\equiv 0$.

Theorem 1.2 If we lose sight of the addition of arbitrary analytic function, $w(z)$ is a bianalytic function, $\left.\frac{\partial w}{\partial \bar{z}}\right|_{\Gamma}=r(t)$, $t\in G$, and the $r(t)$ is known continuous function, then $w(z)$ is unique.

In fact, from (3.1.3) we will have

$$w(z)=\frac{-1}{\pi}\iint_{G}\frac{f(\xi,\eta)}{\zeta-z}d\xi d\eta\equiv T_{G}(f) \tag{3.1.6}$$

If we lose sight of the addition of arbitrary analytic function $\Phi(z)$, here $f(z)=\frac{\partial w}{\partial \bar{z}}$ is an analytic function, obviously, as long as $\left.\frac{\partial w}{\partial \bar{z}}\right|_{\Gamma}=f|_{\Gamma}=r(t)$, $f(z)$ is unique, that is, $w(z)=T_{G}(f)$ is also unique.

For simplicity in the later we lose sight of the addition of arbitrary analytic function $\Phi(z)$ and also will say $w(z)$ is a bianalytic function.

1.2 Fundamental System of Bianalytic Function

We now consider the following function systems:

$$\begin{aligned} w_{2n.0}(z)&=T_{G}(z^{n}),\\ w_{2n+1.0}(z)&=T_{G}(iz^{n}),\end{aligned}\qquad n=0,1,2,\cdots$$

$$\begin{aligned} w_{2n.k}(z) &= T_G[(z-z_k)^{-n}], \\ w_{2n+1.k}(z) &= T_G[i(z-z_k)^{-n}], \end{aligned} \quad n=1,2,\cdots \quad k=1,2,\cdots,m \tag{3.1.7}$$

there z_1, z_2, $\cdots$, z_m are also fixed points in supplementary of complex connected domain for simply connected domain (that is $m=0$) we have only $w_{n,0}(z)$, simply denoted by $w_n(z)$.

It is clear that arbitrary bianalytic function can be expressed by linear combination of function systems (3.1.7) with real coeffocixents and the combination is converge uniformly in G.

We will say function system (3.1.7) is fundamental system of bianalytic function.

Theorem 1.3 Suppose G is circle field: $|z|<R$, $w(z)$ is a biannlytic function in G, and $w(z)$ is continous on $G+\Gamma$, then $w(z)$ can be expressed by uniformly convergent series in G (general Taylor Series).

$$w(z)=\sum_{n=0}^{\infty} C_n w_n(z) \tag{3.1.8}$$

here

$$w_n(z)=w_{n,0}(z)$$

$$C_{2n}(z)=\mathrm{Re}\left(\frac{1}{2\pi i}\int_\Gamma \frac{\partial w/\partial \bar{t}}{t^{n+1}}dt\right), \quad n=0,1,2,\cdots \tag{3.1.9}$$

$$C_{2n+1}(z)=\mathrm{Im}\left(\frac{1}{2\pi i}\int_\Gamma \frac{\partial w/\partial \bar{t}}{t^{n+1}}dt\right)$$

In fact, $\partial w/\partial \bar{z}=f$ is an analytic function in G and it is coutinuous on $G+\Gamma$, so that

$$\frac{\partial w}{\partial \bar{z}}=\sum_{n=0}^{\infty} a_n z^n$$

here

$$a_n=\frac{1}{2\pi i}\int_\Gamma \frac{\partial w/\partial \bar{t}}{t^{n+1}}dt$$, it is convergent uniformly in G.

Further we consider (3.1.6), then

$$w(z)=\sum_{n=0}^{\infty} C_n w_n(z)$$

Obviously

$$C_{2n}=\mathrm{Re}\left(\frac{1}{2\pi i}\int_{\Gamma}\frac{\partial \mathrm{w}/\partial \bar{t}}{t^{n+1}}\mathrm{d}t\right)=\mathrm{Re}a_n,$$
$$n=0,1,2,\cdots$$
$$C_{2n+1}=\mathrm{Im}\left(\frac{1}{2\pi i}\int_{\Gamma}\frac{\partial \mathrm{w}/\partial \bar{t}}{t^{n+1}}\mathrm{d}t\right)=\mathrm{Im}a_n,$$

Theorem 1.4 Suppose G is a circular ring domain $R_1<|z|<R_2$, $\mathrm{w}(z)$ is a bianalytic function in G, and it is a continuous function on $G+\Gamma$, then $\mathrm{w}(z)$ can be expressed by uniformly convergent series in G (general Laurent series).

$$\mathrm{w}(z)=\sum_{n=-\infty}^{\infty}C_n\mathrm{w}_n(z) \tag{3.1.10}$$

where

$$\mathrm{w}_n=\mathrm{w}_{n,0},$$
$$\mathrm{w}_{-2n}=T_G(z^{-n}),\qquad n=1,2,\cdots \tag{3.1.11}$$
$$\mathrm{w}_{-2n+1}=T_G(iz^{-n}),$$

$$C_{2n}=\mathrm{Re}\left(\frac{1}{2\pi i}\int_{\Gamma_2}\frac{\partial \mathrm{w}/\partial \bar{t}}{t^{n+1}}\mathrm{d}t\right),$$
$$n=0,1,2,\cdots,\Gamma_2:|z|=R_2$$
$$C_{2n+1}=\mathrm{Im}\left(\frac{1}{2\pi i}\int_{\Gamma_2}\frac{\partial \mathrm{w}/\partial \bar{t}}{t^{n+1}}\mathrm{d}t\right),$$

$$C_{-2n}=-\mathrm{Re}\left(\frac{1}{2\pi i}\int_{\Gamma_1}\frac{\partial \mathrm{w}}{\partial \bar{t}}t^{n+1}\mathrm{d}t\right),$$
$$n=0,1,2,\cdots,\Gamma_1:|z|=R_1$$
$$C_{-2n+1}=\mathrm{Im}\left(\frac{1}{2\pi i}\int_{\Gamma_1}\frac{\partial \mathrm{w}}{\partial \bar{t}}t^{n+1}\mathrm{d}t\right),$$

For simplicity in the later $\sum_{n=0}^{\infty}C_n\mathrm{w}_n(z)$, and $\sum_{n=-1}^{-\infty}C_n\mathrm{w}_n(z)$ in (3.1.10) will say the bianalytic and the principal part of function $\mathrm{w}(z)$ respectively.

Similarly, we can consider general Taylor series and general Laurent series on $(z-z_0)$.

Chapter Ⅲ Bianalytic Function, Complex Harmonic Function and Their Basic Boundary Value Problems

Due to their wonderful properties, analytic functions possess a lot of applications mechanics and physics-mathematics. For example, when we research the physical fields on plane without source and curl, the theory of analytic functions display their strength forces. But if discuss the physical fields with sources or curls, then this important tool of analytic functions will has no use.

The concept of semi-analytic functions and their properties has suggested and considered. But from the definition of semi-analytic functions, we know there for one equation of $C-R$ equation system have no any requires, thus in applications there will be a large number of difficulies on problem of uniqueness. So that we try to add some control conditions to semi-analytic functions. Therefore we introduce some classes of complex functions, (i. e. bianalytic functions and complex harmonic functions) and prove some theorems for them.

The complex form of $C-R$ equations is

$$\frac{\partial \mathrm{w}}{\partial \bar{z}}=0 \tag{3.2.1}$$

here $\quad \frac{\partial}{\partial \bar{z}}=\left(\frac{\partial}{\partial x}+\mathrm{i}\frac{\partial}{\partial y}\right), \mathrm{w}=u+\mathrm{i}v, z=x+\mathrm{i}y$

In addition we define the derivitives: $\frac{\partial}{\partial z}=\left(\frac{\partial}{\partial x}-\mathrm{i}\frac{\partial}{\partial y}\right)$ and $\Delta=\frac{\partial^2}{\partial \bar{z}\partial z}=\frac{\partial^2}{\partial z\partial \bar{z}}$.

Definition 2. 1 Suppose G is a region on plane and define in a complex function $\mathrm{w}(z)$, in which exists the two order derivitive $\frac{\partial^2 \mathrm{w}}{\partial \bar{z}^2}$ on $\bar{z}$, then we will say it is a bianalytic function, if $\mathrm{w}(z)$ satisfies the following partial diferential equation:

$$\frac{\partial^2 w}{\partial \bar{z}^2}=0, z\in G \tag{3.2.2}$$

We denote the set of all bianalytic functions by $D_2(G)$.

Definition 2.2 Suppose G is a region on plane and define in G a complex function $w(z)$, which exists the two order mixed derivitive $\frac{\partial^2 w}{\partial \bar{z}\partial z}$ then we will say it is a complex harmonic function, if $w(z)$ satisfies the following partial differential equation:

$$\Delta w=\frac{\partial^2 w}{\partial \bar{z}\partial z}=0 \quad (z\in G) \tag{3.2.3}$$

We denote the set of all complex harmonic functions by $H_2(G)$.

2.1 Bianalytic Functions

From equation (3.2.2) we get $\frac{\partial w}{\partial \bar{z}}=f, f(z)=f_1(x,y)+\mathrm{i}f_2(x,y)$, which is an arbitrary analytic function.

Furthermore

$$w(z)=\frac{-1}{\pi}\iint_G \frac{f(\zeta)}{\zeta-z}d\xi d\eta+\Phi(z) \tag{3.2.4}$$

where $\zeta=\xi+\mathrm{i}\eta\in G$ and $\Phi(z)$ is an arbitrary analytic function.

Theorem 2.1 (Uniqueness theorem)

Suppose $w(z)$ is a bianalytic function in G, if

$$\left.\frac{\partial w}{\partial \bar{z}}\right| \partial G=0 \tag{3.2.5}$$

$$w| \partial G=0 \tag{3.2.6}$$

then $w(z)\equiv 0$, $z\in G$, here ∂G is the boundary of G.

Proof: Because $\frac{\partial w}{\partial \bar{z}}=f(z)$ is an analytic function, consequently, from (3.2.5) we will have $\frac{\partial w}{\partial \bar{z}}\equiv 0$, $z\in G$ and from (3.3.6) will have $w(z)\equiv 0$, $z\in G$.

The following theorem holds:

Theorem 2.2 (the first representation formula)

If $w(z) \in D_2(G)$, then

$$w(z) = \frac{-1}{\pi}\iint_G \frac{\frac{\partial w}{\partial \overline{\zeta}}}{\zeta - z} d\xi d\eta + \Phi(z) \tag{3.2.7}$$

here $\zeta = \xi + i\eta \in G$, $\Phi(z)$ is an arbitrary analytic function.

Theorem 2.3 (the second representation formula)

If $w(z) \in D_2(G)$, then

$$w(z) = \overline{z}\varphi(z) + \Phi(z) \tag{3.2.8}$$

here $\varphi(z)$ is an arbitrary analytic function.

Proof: It is clear that $w_1(z) = \overline{z}\varphi(z)$ is a bianalytic function and $\frac{\partial w_1}{\partial \overline{z}} = \varphi(z)$, then $\frac{\partial (w - w_1)}{\partial \overline{z}} = 0$. Furthermore $w(z) = \overline{z}\varphi(z) + \Phi(z)$.

For simplicity in the later we lose sight of the addition of arbitrary analytic function $\Phi(z)$ and also will say $w(z)$ is bianalytic function.

From the second representation formula we know the following theorem holds:

Theorem 2.4 (theorem on zeroes)

If $w(z) \in D_2(G)$, then its zeroes are isolated and the number of zeroes is finite.

Theorem 2.5 (theorem on singularities)

If $w(z) \in D_2(G)$, then the points of singularities are isolated and the number of singularities is also finite.

Theorem 2.6 (Taylor series development)

If $w(z)$ is bianalytic function in disc G, $G: |z| < R$, then in G it holds the following expansion:

$$w(z) = \sum_{k=0}^{+\infty} c_k \overline{z} z^k \tag{3.2.9}$$

where

$$c_k = \frac{1}{2\pi i}\int_{\partial G} \frac{\frac{\partial w}{\partial \overline{t}}}{t^{k+1}} dt, \quad k = 0, 1, \cdots \tag{3.2.10}$$

It is clearly, because $w=\bar{z}\frac{\partial w}{\partial \bar{z}}$, $\frac{\partial w}{\partial \bar{z}}$ is analytic function.

Theorem 2.7 (Laurent series development)

Let $w(z)$ be a bianalytic function in annulus G, $G: R_1<|z|<R_2$, $R_1<R_2<+\infty$, then

$$w(z)=\sum_{k=-\infty}^{+\infty} c_k \bar{z} z^k \tag{3.2.11}$$

where the convergence is absolute and uniform over the annulus $G^* \in G$, G^*:

$r_1<|z|<r_2$, if $R_1<r_1<r_2<R_2$. The coefficients are given by the formula:

$$c_k=\frac{1}{2\pi i}\int_r \frac{\frac{\partial w}{\partial \bar{t}}}{t^{k+1}} dt \tag{3.2.12}$$

where r is the circle $|z|=r$, for any r, $R_1<r<R_2$. Moreover this series is unique.

Proof: Using the second representation formula, we can get the Laurent series development for analytic function $\varphi(z)=\frac{\partial w}{\partial \bar{z}}$. So formula (3.2.11) holds.

Later, series $\sum_{k=0}^{+\infty} c_k \bar{z} z^k$ and $\sum_{k=-1}^{+\infty} c_k \bar{z} z^k$ in (3.2.11) will be also said the analytic and the principal part of function $w(z)$ respectively.

Similarly we use the Laurent expansion to classify isolated singularities.

(1) $z=0$ is a removable singularity if and only if (3.2.11) has no principle part.

(2) $z=0$ is a pole of order m if and only if in principle part of (3.2.11) has only m terms and $C_m \neq 0$.

(3) $z=0$ is an essential singularity if and only if principle part of (3.2.11) has infinite terms.

We can also consider the Taylor and Laurent series development about $(z-z_0)^k$, where z_0 is an arbitrary point in G.

Theorem 2.8 If $w(z) \in D_2(G)$, then its real and imaginary part are all biharmonic functions.

Proof: By definition 2.1 we know $\frac{\partial \mathrm{w}}{\partial \bar{z}^2}=0$. Thus

$$\Delta^2 \mathrm{w}=\frac{\partial^4 \mathrm{w}}{\partial \bar{z}^2 \partial z^2}=0, \mathrm{w}=u+iv,$$

consequently, $\Delta^2 u=0$, $\Delta^2 v=0$.

2.2 Complex Harmonic Functions

From equation (3.2.3) for complex harmonic function $\mathrm{w}(z)$ we have $\frac{\partial \mathrm{w}}{\partial \bar{z}}=\bar{f}$, here $f(z)=f_1(x,y)+\mathrm{i}f_2(x,y)$ is an arbitrary analytic function.

Furthermore

$$\mathrm{w}(z)=\frac{-1}{\pi}\iint_G \frac{\overline{f(\zeta)}}{\zeta - z}\mathrm{d}\xi \mathrm{d}\eta + \Phi(z) \tag{3.2.13}$$

where $\zeta=\xi+\mathrm{i}\eta \in G$, and $\Phi(z)$ is an arbitrary analytic function in G.

Theorem 2.9 (uniqueness theorem)

Suppose $\mathrm{w}(z)$ is a complex harmonic function in G, if

$$\left.\frac{\partial \mathrm{w}}{\partial \bar{z}}\right| \partial G=0 \tag{3.2.14}$$

$$\mathrm{w} | \partial G=0 \tag{3.2.15}$$

then $\mathrm{w}(z)\equiv 0$, $z \in G$, here ∂G is boundary of G.

Proof: $\frac{\partial \mathrm{w}}{\partial \bar{z}}=\overline{f(z)}$, $f(z)$ is analytic function, consequently, from (3.2.14) we have $\frac{\partial \mathrm{w}}{\partial \bar{z}}\equiv 0$, $z \in G$, and from (3.2.15) also have $\mathrm{w}(z)\equiv 0$, $z \in G$.

We consider nonhomogeneous $C-R$ equation:

$$\begin{cases} \frac{\partial u}{\partial x}-\frac{\partial v}{\partial y}=p(x,y) \\ \frac{\partial u}{\partial y}+\frac{\partial v}{\partial x}=0 \end{cases}$$

here $p(x, y)$ is an arbitrary harmonic function. Its complex form is

$$\frac{\partial \mathrm{w}}{\partial \bar{z}}=p(x,y),\mathrm{w}=u+\mathrm{i}v \qquad (3.2.16)$$

whose general solution will be:

$$\begin{aligned}\mathrm{w}(z)&=\frac{-1}{\pi}\iint_G \frac{p(x,y)}{\zeta-z}\mathrm{d}\xi\mathrm{d}\eta+\Phi(z)\\&=\frac{-1}{\pi}\iint_G \frac{\varphi(\zeta)}{\zeta-z}\mathrm{d}\xi\mathrm{d}\eta+\frac{-1}{\pi}\iint_G \frac{\overline{\varphi(\zeta)}}{\zeta-z}\mathrm{d}\xi\mathrm{d}\eta+\Phi(z)\end{aligned} \qquad (3.2.17)$$

where $\varphi(z)$ is an arbitrary analytic function, thus

$$\mathrm{w}_1(z)=\frac{-1}{\pi}\iint_G \frac{\varphi(\zeta)}{\zeta-z}\mathrm{d}\xi\mathrm{d}\eta$$

is a bianalytic function and

$$\mathrm{w}_2(z)=\frac{-1}{\pi}\iint_G \frac{\overline{\varphi(\zeta)}}{\zeta-z}\mathrm{d}\xi\mathrm{d}\eta$$

is a complex harmonic function.

By theorem 2.1 and theorem 2.8 we know that for (3.2.17) also holds the theorem uniqueness. Therefore, if add to semi-analytic functions the control condition that $p(x, y)$ in (3.2.16) is a harmonic function, then have the aid of bianalytic and complex harmonic functions we already can consider the physical fields with sources or curls.

2.3 Basic Boundary Value Problems

First, we will discuss boundary value problems of bianalytic function (Question A).

For complex equation:

$$\frac{\partial \mathrm{w}}{\partial \bar{z}^2}=0 \quad (z\in G) \qquad (3.2.18)$$

get $\mathrm{w}(z)$, which satisfies following conditions:

$$\left.\frac{\partial \mathrm{w}}{\partial \bar{z}}\right|\partial G=g_1(t),\mathrm{w}|\partial G=g_2(t) \quad (t\in\partial G) \qquad (3.2.19)$$

here $g_1(t)$ and $g_2(t)$ are given and continous on ∂G.

Theorem 2. 10 Question A is solvable and its solution is unique.

Proof: Because $\frac{\partial w}{\partial \bar{z}}$ is an analytic function, from (3. 2. 19) we know that $\frac{\partial w}{\partial \bar{z}}=\varphi(z), z\in G$ be defined uniquely, further, $w_1(z)=\frac{-1}{\pi}\iint_G \frac{\varphi(\zeta)}{\zeta-z}d\xi d\eta$ be also defined uniquely, consequently from the first representation formula, we have

$$w(z)=w_1(z)+\Phi(z)=\frac{-1}{\pi}\iint_G \frac{\varphi(\zeta)}{\zeta-z}d\xi d\eta+\Phi(z)$$

here $w_1(z)$ is defined uniquely, $\Phi(z)$ is an arbitary analytic function.

Further, from (3. 2. 19) we have

$$w(z)|\partial G=w_1(z)|\partial G+\Phi(z)|\partial G=g_2(t)$$

here

$$w_1(z)\mid \partial G=\frac{-1}{\pi}\iint_G \frac{\varphi(\zeta)}{\zeta-z}d\xi d\eta=w_1(t)$$

is a defined known function, so that question A can be translated boundary value problems.

$$\Phi(z)|\partial G=g_2(t)-w_1(t)=r(t) \qquad (3. 2. 20)$$

and from uniqueness theorem of analytic function we have unique solution of (3. 2. 20).

Similar we can discuss boundary value problems of complex equation (Question B).

For complex equation:

$$\frac{\partial^2 w}{\partial \bar{z}\partial z}=0 \quad (z\in G) \qquad (3. 2. 21)$$

get $w(z)$, which satisfies (3. 2. 21) and

$$\left.\frac{\partial w}{\partial \bar{z}}\right| \partial G=g_1(t), w|\partial G=g_2(t) \quad (t\in \partial G) \qquad (3. 2. 22)$$

here $g_1(t)$ and $g_2(t)$ are known and continous functions on ∂G.

Theorem 2. 11 Question B is solvable and its solution is unique.

Proof: similar to proof of theorem 2. 10, from (3. 2. 22) consider $\left.\overline{\frac{\partial w}{\partial z}}\right| \partial G=\overline{g_1(t)}$, and $\overline{\frac{\partial w}{\partial z}}$ is an analytic function, so that $\overline{f(z)}$ can be defined uniquely. Finaly, for semi-analytic function that be add to some control conditions can be also discussed corresponding boundary value problems. Further, for $w|\partial G=g_2(t)$, $\Phi(z)$ can be defined uniquely.

Interpretation of Semi - analytic Function and Conjugate Analytic Function

In 1983, Professor Wang Jianding first proposed a semi - analytic function in the world. The analytic function is a solution that satisfies the (C - R) equation system, while the semi - analytic function is a solution that only needs to satisfy one equation in the (C - R) equation system. Obviously, the analytic function must be a semi - analytic function, and the semi - analytic function is not necessarily an analytic function.

In 1988, Professor Wang proposed and systematically established the conjugate analytic function theory for the first time in the world. "Conjugate" was studied as a mathematical symbol in middle school. To put it simply, "conjugation" means "symmetry". But "conjugate analytic function" is a completely new function class. As we all know, the derivative integral of a complex variable function is derived in the form of a derivative integral of a real variable function. The value of analytic functions lies in its important applications in electric fields, magnetic fields, fluid mechanics, and elastic mechanics. But after careful study, it is found that all the above applications are direct applications of conjugate analytic functions, not analytic functions. Conjugate derivatives and conjugate integrals have clear and direct physical and mechanical meanings (analytical functions do not have).

Conjugate analytic function and analytic function are completely symmetrical, which makes complex variable function perfect. As we all know, symmetry is a general beauty in science. In fact, with these two smallest functions, $w=z$ (analytic function), $w=\bar{z}$ (conjugate function), the different combinations of them form variable complex functions.

References

[1] I. N. VEKUA, Generalized Analytic Functions [M]. New York Martino Fine Books, 2014.

[2] L. BERS, Theory of Pseudo-analytic Functions [M]. New York: New York University, 1952.

[3] LI Zhengdao, Mathematical Methods in Physics [M]. Nanjing: Jiangsu Science and Technology Publishing House, 1980 (in Chinese).

[4] ZHANG Zongsui, Electrodynamics and Special Theory of Relativity [M]. Beijing: Science Publishing House, 1957 (in Chinese).

[5] WANG Longfu, Elastic Theory [M]. Beijing: Science Publishing House, 1978 (in Chinese).

[6] WANG Jianding. Semi-analytic Function [J]. Journal of Beijing Polytechnic University, 1983 (3): 45-51.

[7] WANG Jianding. Improvement of Two Theorems of Semi-analytic Function [J]. Journal of Beijing Polytechnic University, 1984 (4): 127.

[8] WANG Jianding. Existence Theorems of Semi-analytic Function and Mutual Translations Between First and Second Class Semi-analytic Functions [J]. Journal of Beijing Polytechnic University, 1984 (4): 127.

[9] WANG Jianding. Semi-analytic Extension [J]. Journal of Beijing Polytechnic University, 1984 (4): 128.

[10] WANG Jianding. Several Equivalence Theorems of Semi-analytic Function [J]. Journal of Beijing Polytechnic University, 1986 (1): 22.

[11] WANG Jianding. Decomposition Theorem of Complex Function [J]. Journal of Beijing Polytechnic University, 1986 (2): 6.

[12] WANG Jianding. Semi-analytic Representations of Some Continuous Complex Functions [J]. Journal of Beijing Polytechnic University, 1988 (4): 55-57.

[13] Fan Dajun. Methematic Elatrodynamics [M]. Beijing: New Era Publishing House, 1983 (in Chinese).

[14] WANG Jianding. A new Proof of Decomposition of Complex Function and lts Uniqueness [J]. Journal of Beijing Polytechnic University, 1991 (1): 60.

[15] WANG Jianding. Semi-analytic Function of Integral Form [J]. Journal of Beijiing Polytechnic University, 1991 (4): 16-17.

[16] WANG Jianding. Integral Form of Semi-analytic Function [J]. Journal of Beijing Pol-

ytechnic University, 1991 (4): 18 - 20.

[17] ZHU Ruzhen, ZHU Ying. Improvement on Decomposition Theorems of Complex Functions [J]. Journal of Beijing Polytechnic University, 1994 (3): 119 - 120.

[18] WANG Jianding. Semi-analytic Function, Conjugate analytic Function and Their Application in Mechanics, Advances in Mechanics, 1997 (2): 257 - 263.

[19] ZHAO Zhen. Some Properties of Bianalytic Function [J]. Journal of Sichuan Normal University (Natural Science), 1994 (2): 114 - 116.

[20] ZHAO Zhen. Bianalytic Functions, Complex Harmonic Functions and Their Basic Boundary Value Problems [J]. Journal of Beijing Normal University (Natural Science), 1995 (2): 175 - 179.

[21] ZHAO Zhen. Bianalytic Function and Applicntions, Proceedings of the Second Asian Methematic Conference [R]. Thailand, 1995.

[22] WANG Jianding. Semi-analytic Function and Conjugate Analytic Function [M]. Beijing: Beijing Polytechnic University Publishing House, 1988.